Copernicus Books

Sparking Curiosity and Explaining the World

Drawing inspiration from their Renaissance namesake, Copernicus books revolve around scientific curiosity and discovery. Authored by experts from around the world, our books strive to break down barriers and make scientific knowledge more accessible to the public, tackling modern concepts and technologies in a nontechnical and engaging way. Copernicus books are always written with the lay reader in mind, offering introductory forays into different fields to show how the world of science is transforming our daily lives. From astronomy to medicine, business to biology, you will find herein an enriching collection of literature that answers your questions and inspires you to ask even more.

Giovanni Samaey • Joos P. L. Vandewalle

The Invisible Power of Mathematics

The Pervasive Impact of Mathematical Engineering in Everyday Life

 Springer

Giovanni Samaey
Department of Computer Science
Katholieke Universiteit Leuven
Heverlee, Belgium

Joos P. L. Vandewalle
Faculty of Engineering
Department of Electrical Engineering
Katholieke Universiteit Leuven
Heverlee, Belgium

The Work already has been published in 2016 in Dutch language by Pelckmans Pro with the following title: X-factor, 20 verhalen over de onzichtbare kracht van wiskunde. Therefore, notwithstanding the above, the rights for translation of the Work into Dutch language and for publication of a Dutch edition are excluded from the rights granted to the Publisher under this Agreement.

The Author ensures that they have retained or retrieved from the former publisher (in writing) the rights granted under this Agreement.

ISSN 2731-8982   ISSN 2731-8990   (electronic)
Copernicus Books
ISBN 978-1-0716-2775-4   ISBN 978-1-0716-2776-1   (eBook)
https://doi.org/10.1007/978-1-0716-2776-1

Foreword

This is a book I would love to have had on my own bookshelf growing up, although of course some of the mathematics discussed here was not yet developed at the time. But a considerable part existed then already, and I would have been fascinated to read about the mathematical ideas and techniques essential to understanding the world around us. I am therefore grateful that Giovanni Samaey and Joos Vandewalle wrote this book, vividly describing mathematical explorations for a wide audience. I hope it will help to convince many young people and adults of the versatility of mathematics, and will prompt them to recognize within themselves the mathematical curiosity that can be found living inside almost everyone, once one starts digging. Just as everyone can enjoy sports, even if only a few become an Olympic champion, so too can everyone enjoy math thinking, even if it doesn't make you a top mathematician. And it is a lot easier for you to earn a good living with math than with sports! And just as it is healthier to incorporate physical activity into your life, your life can also be happier if you cultivate a sense of wonder and the desire to understand how things work, such as the mechanisms of public key cryptography, image compression for digital cinema or the organization of global freight shipping, to name but a few examples, not even all those are discussed in this book. Maybe in a sequel...

Enjoy reading!

Department of Mathematics
Duke University, Durham, NC, USA

Ingrid Daubechies

Acknowledgment

This book deals with the reasons why we engage in math. It contains many of the societal challenges that motivate us to dive deep into mathematics as well as lots of anecdotes we like to share at parties and with friends. It is about stuff that we think should be of interest to everyone at some level, mathematician or not. Yet writing a book like this is not the easiest task for an academic. Not because the content would be too difficult (because it isn't), but because we usually spend our time explaining in excruciating mathematical detail to our colleagues what we have done, instead of talking about the reasons we do it with our friends.

We are therefore very grateful that Helena Slechten, our Flemish publisher, contacted us quite some years ago around the idea of 'doing something about the usefulness of mathematics'. In between our other activities, she especially allowed us the important maturing time for writing. Our conversations with Helena have shaped this book into the form it is today, and of which we are proud. (Whether that pride is justified, we leave to your judgment.) We thank Nancy Derboven for her great contribution to the design and finishing of this book. We are also very grateful for all the help we received in collecting the content, in particular from Joris De Schutter (KULeuven), Herman Van der Auweraer (LMS / Siemens), Ann Dooms (VUB), Isabelle Thomas and Benjamin Wayens (Brussels Studies).

Writing about mathematics in today's and tomorrow's world is not without risk. It doesn't take much to quickly get lost in a technical detail, when it is really only necessary to convey the crucial ideas. Fortunately, we could count on a motivated group of guinea pigs to read our first versions: Sabine Bosteels, Liesje Knaepen, Ward Melis, Bert Mortier, Keith Myerscough, Kobe

Nachtergaele, Anne-Sophie Putseys, Pieterjan Robbe, Bert Seghers and Pieter Van Nuffel, thanks!

Finally, writing this book also placed a certain burden on our private life. To our respective spouses Valérie and Rita and our children and grandchildren, Lukas and Emma, Patrick & Els, Johan & Leen, Ellen & Luc, Mats, Amber & Lana, Janne, Sien, & Nel: thank you for the moral support, sorry, and hopefully there is now some more time for you.

May 2016

The English version also took a while to mature and come to life. In fact, when we talked proudly about our Dutch book to our colleagues at coffee breaks in international meetings, many told us that they were not aware of such a book in the English or international literature. Moreover, they stressed that at an international scale there is also a need for such a book with stories that prove the wide applicability of mathematics. Stimulated by these encouragements, we engaged with Springer for the English version. We are thankful for the efforts of Elizabeth Loew for her support and intense communication.

January 2021

Giovanni Samaey and Joos Vandewalle

Introduction

Whenever someone finds out that we have turned 'mathematics' into our profession, two remarks always return – usually politely disguised as a question. The first reaction is often: 'To make math your job, you have to be really smart (and a little bit of a weirdo)'. The second one is: 'Does all that mathematics really serve any purpose?' Our answers are simple: 'Mathematics requires some training, but is certainly not just for a small group of geniuses'. And 'Yes, mathematics serves more than you can imagine. You can help a lot of people with it.'

This book deals in the first place with the second question. Through 20 stories, which can be read separately, we show how mathematics helps to shape the world around us. This is explicitly *not a mathematics book*, but a book about the role that mathematics plays in devising the creative solutions the world needs. In the first part, *Domestic, Garden and Kitchen Mathematics,* we deal with themes from everyday life, whether it's online shopping, playing board games, the operation of a hearing aid or reading an opinion poll in the newspaper. In the second part, *Mathematics in the Workplace,* we look at professional life. We not only show how mathematics plays a role for engineers making new cars or working on renewable energy, but also how art historians use mathematics to check the authenticity of paintings and carry out restorations. Finally, in the third part *Mathematics for Tomorrow's Society,* we look at some of the major challenges facing society, such as global warming, the ageing population, the spread of a pandemic, mobility or the pursuit of world peace. The themes are interspersed with short intermezzos that bring out a remarkable figure or fun facts.

Each chapter stands alone and deals with a specific theme. Each time, we conclude from everyday reality that we have a problem that we can't really solve unless we use some math. Here and there, a mathematical detail, or a historical anecdote, is worked out in more detail in a separate box. Again, we only explain the idea behind a certain mathematical technique, and consistently avoid all technical details. By the way, those pieces can be skipped perfectly (although we would regret it if that would happen systematically). We are not concerned with the formulas – there are hardly any in the book either – but with the idea, the insight that made a solution possible. To make you, the reader, think: 'Hey, what an original and beautiful solution they have come up with'. In this way, we want to create a sense of *wonder* about the hidden power and omnipresence of mathematics in our daily lives, in our jobs, and for the challenges of the future.

This brings us back to the first question: 'Do you really have to be a superhuman genius to be involved in mathematics?' Fortunately, the answer to that question is negative. Mathematics is far too important to be left to mathematicians alone! The British mathematician Ian Stewart formulated it as follows: suppose we would put a small red sticker on each object if mathematics was used for its development. Then we wouldn't be able to look anywhere without seeing dozens of red stickers: no TV, no car, no hospital, no WhatsApp, Zoom or Skype meeting or film can avoid the sticker. Even the vegetables we eat have been developed through hybridization programs that were mathematically controlled since the beginning of the twentieth century. It is no coincidence that genetics, as one of the first biological domains, was described entirely mathematically. It is impossible to attribute this gigantic progress to a handful of brilliant mathematicians. There are simply too few of them!

However, public opinion is often dominated by the idea that mathematics should be primarily 'beautiful'. That mathematics is 'an end in itself'. That mathematics is a 'cultural element' 'with its own history and methods'. That mathematics is primarily an 'intellectual exercise', a test for the human mind. And there certainly are a lot of mathematicians to be found who deal with abstract, 'pure' mathematics. The fact that some of them exhibit bizarre social behaviour sometimes draws the attention of the public even further away from the practical usefulness of mathematics. For instance, the Russian Grigori Perelman is perhaps better known for his refusal of both the Fields Medal – one of the highest awards in the world – and the Clay Prize of $1 million than for his proof of the Poincaré conjecture. Yet such strange characters are very exceptional, even within the pure mathematics world.

Moreover, this whole group of 'pure mathematicians' is only a fraction of the people who come into professional contact with mathematics themselves. As the examples in this book will show, mathematics has an important impact in almost all aspects of our daily lives. Politicians and journalists, psychologists and doctors, historians and art experts, climatologists, epidemiologists, biologists, computer scientists and engineers; the list of professions in which mathematics plays a role is almost endless. Even a book like ours can only show the tip of the iceberg. Of course, we won't claim that you must be a mathematical expert for all these professions (that would just not be true), but a certain insight into what mathematics can (and cannot) do is not a superfluous luxury.

Conversely, unworldliness is also not a requirement for a good mathematician. In order to make mathematics useful for solving technical or social problems, it is very important to be able to describe a 'real' problem very precisely so that it can be converted into a 'mathematical' problem that can then be solved by (applied) mathematicians. In practice, mathematicians often work in teams, together with experts in other fields, and from their concrete knowledge, reach solutions that other team members might not suddenly think of. The examples in this book illustrate this in a very diverse way: they all start from a concrete question that concerns an individual, a company, or society as a whole, and translate this question into a mathematical problem. In this way, we can see mathematics as a 'tool', as a means to an end. So, it is a misconception to think that 'helping people' and 'studying mathematics' are mutually exclusive! With mathematics, you often help a team of colleagues develop new technology, and that technology itself often helps entire generations of people.

Wherever possible, we show what mathematics has already achieved, but just as often we will have to acknowledge that there are limits to what we can do now. Sometimes, we can push those limits by applying more existing mathematics. But at least as often we conclude that a certain piece of mathematics is missing, and that extra mathematical work is needed to come to a solution. Because the role of mathematics in our society is steadily increasing, it happens more and more often. This may sound paradoxical – because the mathematical knowledge itself is also increasing – but this only illustrates the ambition of humankind: the more we can do, the more we want to do! This observation leads to a growing importance of mathematics in science in higher education, especially for engineers. It is no wonder that many universities, including our own KU Leuven in Belgium, are creating specialized Master's degrees in Mathematical Engineering that specifically trains engineers to use a wide range of mathematical techniques to solve technical and social problems.

This book is for anyone who wonders what role mathematics plays in our society. For anyone who looks back at their school days and wonders what mathematics was for. For teachers who want to tell their pupils this during their school days, and for pupils who want to decide for themselves. For young people who are wondering which way they want to go with their lives, and for everyone who wants to help them with that. Have fun with it.

Giovanni Samaey and Joos Vandewalle, May 2016, and January 2023 for the English version.

Contents

Part I   Mathematics in Everyday Life

1   Worry-Free Online Shopping.   3

2   Reach Everyone, Anywhere, in Just Six Steps   11

    Teapots, Ponytails, and an Unexpected Reward   20

3   Are the Polls Right?   21

4   Cochlear Implants Help the Deaf Hear Again   31

    Perception is Relative, and Sometimes 100 + 100 = 106.   37

5   Google PageRank, or the Needle in the Haystack.   39

6   Fun and Game Theory   47

Part II   Mathematics in the Workplace

7   High Winds on High Mills   55

8   Reducing Factory Noise   61

    The Perfect Engine Sound   66

9   Virtual Architecture and Engineering   69

10   The Art of Forgery   77

11   The Right Bike in the Right Place   83

    Robert Bridson, Mathematician and Oscar Winner   89

12   Can Computers Detect Fraud? (And Do We
    Want Them To?)   91

13   Industrial Fortune Tellers Predict Profit   101

Part III   Mathematics for Tomorrow's Society

14   Do Smart Girls Stay Single Longer?   111

15   What If There's More Data Than Storage?   119

    Stelarc, the Man with the Third Arm   127

16   Driving Without a Driver   129

17   How Vulnerable Are Our Banking Systems?   137

18   Can We Predict Climate Change?   145

    On Rubber Ducks and Other Toys   151

19   War and Peace.   153

20   Pandemics: From Ebola and HIV to Bioterrorism
     and the Coronavirus.   159

Bibliographic References   167

Index   171

Part I

Mathematics in Everyday Life

Chapter 1
Worry-Free Online Shopping

The convenience of online shopping and banking has completely changed our consumer behaviour. From behind our computer, with only a few clicks, we can carry out any financial transaction imaginable, day and night. But how can all this online activity be carried out safely? How do we know that we are on the site we intended to visit and not on a fraudulent copy? And how can the bank be sure that we are who we portray to be and that the transaction we intended to carry out has been passed on to them correctly?

What's the problem? Online banking has existed since 1994. Websites such as Amazon and eBay saw the light of day in 1995. Since then, internet commerce has become even more important. More than half of the European population uses the internet to do their banking. Figures from the Netherlands show that in 2014 online shopping accounted for 22% of all purchases (and even 66% of services purchased). The global pandemic of 2020 has only accelerated this evolution. We use the internet for almost all of our personal communication, for example via email or Whatsapp. We store important documents *in the cloud* with services like Dropbox. We buy apps from Apple's App Store or Google's Play Store. We buy books, train tickets and concert tickets online. We book travel online. And even all our bank accounts are conveniently accessible via the Internet. And we never see who we are in contact with. It is clear that this way of life can only survive if the system to make transactions safe is watertight and trusted.

© The Author(s), under exclusive license to Springer Science+Business Media, LLC, part of Springer Nature 2022
G. Samaey, J. P. L. Vandewalle, *The Invisible Power of Mathematics*, Copernicus Books, https://doi.org/10.1007/978-1-0716-2776-1_1

Know who you are talking to Virtually all online services try to ensure your identity by asking you to enter your password or credit card number when you log in. This is of course already the first problem: if someone is eavesdropping on your internet connection at the time your password passes by, they will be able to use it immediately to pretend to be you. So that password is not just forwarded, but first 'encrypted' or enciphered, expensive words for 'converted to coded language'. Websites that encrypt passwords before sending them over the Internet can be recognised by the prefix 'https://'. The added 's' stands for 'secure' and means that the standard SSL protocol is used for encryption. (Just like the 'protocol' is used at the royal court or in diplomatic circles to indicate the set of rules that everyone has to follow, the SSL protocol is an agreed set of rules that all secured internet communication must follow.) The SSL protocol regulates two important matters. Firstly, it ensures that your data is forwarded to the intended recipient and not to someone who mimics the recipient. 'Phishing', where a swindler pretends to be your bank, is thus excluded. On top of that, the algorithm encrypts the message so that only sender and receiver can read it. The code of encryption is agreed between sender and receiver at the beginning of the communication. Moreover, this is done in such a way that nobody can intercept the message along the way, even if the agreements between sender and receiver themselves are intercepted. But how can that be done?

A box and a key The complete list of rules in the SSL protocol is quite extensive. Since we want to concentrate on the mathematics under the hood, we can only give a glimpse. We look at one specific aspect, namely that of encrypting the message between sender and receiver. In cryptography, the names Alice (for the sender) and Bob (for the receiver) are usually used. A message is always sent from Alice (A) to Bob (B). In the world of online communication, a key is a number or a set of numbers. When encrypted with public–private keys, each recipient creates a key that consists of two parts: a public part that everyone can see and a private part that is known only by the recipient. The fact that the public part of the key does not have to be hidden has far-reaching consequences: it eliminates the need to agree on the keys with each individual sender. When a sender wants to send a message, all he (or she, in the case of Alice) has to do is look up the public key of the receiver (via the Internet) and the communication can be encrypted. Because both parties have a different piece of information, we call this type of encryption asymmetric. (There are also symmetric methods, where sender and receiver use the same key, but these can only be used if the key itself is exchanged in a secure way. This problem is avoided with asymmetric encryption.)

The principle can best be compared with a padlock. The receiver (Bob) makes an open padlock available in the public space. Bob asks whoever has a message for him to put it in a box and to close the box afterwards with his padlock. When Alice does that, she can then send her closed box with the message to Bob, and the box will be locked during transport. Because only Bob has the corresponding key, he is the only one who can open the box again upon arrival and he is therefore the only one who can decipher the message. Any villain on the internet can find or see Bob's open padlock, but can't really do anything with it. And when the locked box passes by, an open padlock will be of no use to open the box. Of course, we don't send physical padlocks over the internet. That role is taken over by numbers in the algorithm. The digital counterpart of the open padlock is the public part of the encryption key. The private part is the padlock key, which Bob has to keep safe. In 1977 a well-known secure algorithm was developed by Ron Rivest, Adi Shamir and Len Adelman. It was called RSA and bases its public key on the product of two

---

Large semiprime numbers as keys

To understand the idea of public–private encryption, we first look at a simple case. This case will be easy to crack – and therefore not really useful for encryption – but it gives a start and it lays down the main idea. We place ourselves in the role of the recipient (Bob) who wants to make a public and private key. Bob starts by choosing a prime number, which we call M. In practice, M is quite large, but here we take a small prime number ($M = 23$, to be specific), so that we are able to perform a few calculations by hand. We agree that we will now consider all the integers from 1 to $M-1$ as possible messages. These are (not coincidentally) all numbers that can be obtained as remainder when we divide a large number by M, excluding zero.

Bob puts the number M at the *disposal of* the transmitter (Alice), together with a second number, P. The numbers M and P together form the public key. To illustrate, we choose $P = 9$. Bob now gives Alice the following assignment: don't send me your message directly, but do the following:

- raise your message to the power P,
- divide the result by M and remember the remainder,
- send me this remainder.

We'll make this procedure concrete with an example. Alice wants to send the number 7. She listens carefully to Bob, who has given her the numbers M and P, and first performs a few calculations. She calculates as follows:

$$7^9 = 40353607 = 1754504 \times 23 + 15$$

(continued)

and decides she has to send out the number 15 (the remainder of 40 353 607 when divided by 23) to Bob. Bob now has a secret piece of information, namely the number $G = 5$. This number G is his private key. (We will tell you where G comes from in a moment.) Bob takes the message that came in (in this case, 15), and does the following with it:

- he raises the message to the power G,
- he also calculates the remainder after dividing by $M = 23$:

$$15^5 = 759375 = 33015 \times 23 + 7$$

Bob decides that the original message must have been the number 7.

That the trick above always produces the right result is guaranteed by the so-called 'little theorem of Fermat', a purely theoretical result from the first half of the seventeenth century. This theorem guarantees that, for each public key M and P, we can calculate a private key G for which the algorithm above works perfectly – except when we choose P really badly. Unfortunately, the theorem also tells you how to calculate G if you know M and P. So that's not really useful. The encryption can only be safe when strangers have a hard time guessing the secret number G used by the receiver, and that's not the case yet.

Fortunately, there is a more general theorem that uses 'semiprimes', numbers that are not prime numbers themselves, but the product of two prime numbers. When the number M used above is not a prime number itself, but a semiprime number, then we can only calculate the number G easily when we know the decomposition of M in its prime factors. Let us call these prime factors Q and R here. Therefore, we have $M = QR$ with Q and R *prime* numbers. Calculating Q and R from M turns out to be a very difficult problem, because the necessary calculation time – with the best existing methods – increases exponentially with the length of M. At the same time, Bob can very easily set up his code, because he knows perfectly which two prime numbers Q and R he started from. So, it is very important that Bob only passes on the number M to Alice, and keeps Q and R secret. If Bob can do that, the above algorithm is a good encryption. Even if computers would get much faster, it suffices to just take a larger number M. Nowadays, it is recommended to use a number M that consists of about six hundred decimal digits.

large prime numbers. Because the algorithm is difficult to crack, it is used all over the world. One of the most important computer programs based on the algorithm is aptly called PGP, which euphemistically stands for 'Pretty Good Privacy'.

How do you know for sure who the recipient is? A large part of security becomes worthless when we as a sender cannot be sure of the identity of the receiver. Therefore, in practice, we need to be able to verify whether the public key we receive really belongs to the intended receiver. With the RSA protocol, there is only one job left: ensuring that the public key really belongs to the receiver. The SSL protocol in your browser regulates this with a certificate that is 'digitally signed' by a reliable third party, for example, a company such as Symantec that makes antivirus software. In cryptography, it is customary to call this third party 'Trent'. (That may be a somewhat contrived name, compared to Alice and Bob, but it starts with the same letters as the word 'trustworthy'). Digital signature works in *just the opposite way* of encryption: Trent sends the transmitter (Alice) a certificate that guarantees the authenticity of the receiver (Bob). This certificate is encrypted with Trent's *private* key. On top of that encrypted certificate, Alice also receives the decrypted certificate. When Alice now decrypts the encrypted message with Trent's *public* key and it corresponds to the decrypted certificate she received, Alice is sure it was really encrypted with Trent's private key. In this way, Trent guarantees Bob's identity.

So how can the NSA intercept your data? Because of all these precautions, no one can intercept encrypted information on public networks, for example, in stations or coffee bars. Yet security services, such as the American NSA, systematically try to read encrypted messages. They often succeed in doing so. Are there holes in the security? The answer is nuanced. First and foremost – in some justified cases – courts can oblige individuals or companies to disclose the private part of their key to the NSA. In addition, of course, there can also be flaws in the security software, even if the method is mathematically perfect. The NSA spends a lot of time and money on detecting such errors and likes to exploit them. Moreover, it turns out that the NSA used a lot of persuasion within the large-scale espionage program PRISM to encourage software companies to deliberately introduce very subtle errors in their programs, which could then be exploited by the NSA.

So how does a Digipass work? When it comes to accessing your bank accounts, the above security is probably not sufficiently reassuring. Extra caution is certainly not a luxury here. For our banking transactions, most people use a Digipass, a small device at home that has the task of generating passwords for one-time use. These passwords can only be used for a maximum of one minute after they have been generated and are of course forwarded with the encryption described above. If a villain is already trying to crack the encryption, you will have a new password by the time he has managed to crack the previous one.

To generate that one-time password, a number of calculations are performed, both in the Digipass and on the bank's server. These calculations are the same for everyone, but they use information that is unique to the transaction. Combining three pieces of information is sufficient. The first ingredient is the identity of the customer. This identity can be verified in two ways, and there are two types of Digipass. Some banks distribute Digipasses that only work when inserting your bankcard. Then, the information on the bankcard can be used to identify you. All Digipasses are identical. You can then borrow a Digipass from a friend or colleague, but you must have your bankcard with you (and of course know the PIN code of the bank card). The alternative is an individual Digipass, in which the identity of the customer is directly encoded and which you can use without the bankcard. If your bank uses this system, you as a user must keep the Digipass as secure as your bankcard itself, which is far from obvious.

The two remaining pieces of information depend on the specific transaction. The first is the time of day. We need to know that because a different password is generated every minute. The second is a 'challenge': a number that depends on the specific transaction, such as the amount or account number of the recipient. On the basis of these three ingredients, the Digipass then calculates a one-time password, which we call the 'response'. Because the server itself also has all the data to calculate the 'response', it can verify the correctness of the one-time password. Not every way to calculate the 'response' is sensible: it is important to organise the calculations in such a way that it is difficult for outsiders to predict the one-time password, even when they know the 'challenge' the bank sends and collect a lot of 'responses' that were given for earlier 'challenges'. This is because the entire security system is of little use if a possible villain can take the time to crack old passwords and predict new ones on the basis of his earlier observations.

The only real disadvantage of electronic banking compared to cash is that the bank always knows the two parties in a transaction. Cash is more anonymous. That has many advantages when you are engaged in illegal activities – such as tax evasion, drugs, human trafficking or terrorism. But most people prefer to also pay for perfectly legal but embarrassing expenses anonymously. And of course, there are fundamental reasons to want to keep your financial situation private. That there is interest in exploiting as much of your financial information as possible is clear. The documents that Edward Snowden leaked in 2013, for example, show that everyone's financial transactions were monitored on a large scale by the NSA. And in 2014, the Dutch bank ING announced that they wanted to use their customers' payment details for personalised advertising.

It would therefore be interesting to make electronic payments without anyone being able to identify the two parties involved in the transaction. Fortunately, this is possible with the encryption techniques we have already seen. The only thing the bank has to do is to provide a communal 'jar' in which money can be put and from which money can be taken. Suppose a sender wants to send 100 euros to a receiver without revealing the receiver's identity. The sender then sends an (encrypted) payment of 100 euros to 'the jar' and at the same time (not via the bank) a digitally signed certificate to the receiver, which entitles him to withdraw 100 euros from that jar. If there are enough transactions to and from the pot, it becomes much more difficult for the bank to trace your payments. Still, because a judge can break the encryption by asking for the private keys, the detection of criminal activity is possible.

And can we completely bypass the banks? Some go a step further, and create a new digital currency that allows us to pay without the intervention of the banks. The best-known example is Bitcoin, which came online in 2009. (Quite a few digital currencies also called cryptocurrencies have been created since then, with varying degrees of success.) Bitcoin bypasses the banks, and therefore eliminates the fees charged by banks. The role of the bank is entirely controlled by software on the computers of Bitcoin users. All those users are connected via the Internet. The task of checking the authenticity of sender and receiver is taken over by the entire network of those users. It is a bit like making a payment in cash only in front of witnesses when you can't trust the other party. Bitcoin is not completely anonymous either, but it takes a little more effort to find out the identity of the users. Of course, digital currencies also have a number of problems. For example, without appropriate legislation, payments in Bitcoin can easily serve to evade taxes. Because there is no real monetary value for the virtual currency, there is a risk of inflation. And speculation by rogue users is also a real danger.

Online shopping and payment have changed our lives dramatically. When the system works, we can consume 24 hours a day, 7 days a week – and everyone at the same time – at a furious pace. Payments are secure and happen almost instantly. The entire electronic payment infrastructure works so flawlessly that we only dwell on this technological miracle when it fails, as happened, for example, in Belgium in the busy run-up to Christmas in 2013 and 2014. In the 3 hours of the breakdown, more than 1 million transactions turned out not to have been carried out correctly. Still, the extensive technical possibilities of cryptography automatically raise questions of a completely different order. The public debate on the legal, ethical and social implications of this technology is so fundamental that it requires sound technical prior knowledge.

Chapter 2
Reach Everyone, Anywhere, in Just Six Steps

How does information spread so fast on social media? Can we get a better understanding of populations and their relationships by studying the structure of social networks? And how do we balance our friendships?

What's the problem? In most social networks, such as Facebook, Google Plus or LinkedIn, people are connected by referring to each other as friends or colleagues. With Twitter, random users can follow each other without the need for a mutual connection. Often, these social networking sites suggest potential acquaintances to expand your network. They do this not only to be sympathetic and helpful but also, for example, because it can be interesting for advertisers: the information you share is then seen by more people. Moreover, the lists of people you may know are often eerily accurate. How and why do social networks do that? How do they look for 'important' people in such a network? How do they find groups or cliques? And how do these structures influence the dissemination of information about the network?

Starting with a figure To tackle these types of questions in a systematic way, we start by representing the network graphically, as was done on Fig. 2.1. Each person then becomes a node in a network, and each friendship between two persons becomes a connection (link). A non-symmetrical link (such as 'following' on Twitter) can be represented graphically by drawing an arrow on the connection. On social networks, each user only has links with a limited number of other users. Facebook now has more than 1 billion users, while an average user has about 200 friends. What's more, friends of your friends are often friends of each other. Facebook's social network therefore consists of a number of groups (clusters) of shared friends with many mutual connections. Most people naturally have friends in different contexts (such as old school

© The Author(s), under exclusive license to Springer Science+Business Media, LLC, part of Springer Nature 2022
G. Samaey, J. P. L. Vandewalle, *The Invisible Power of Mathematics*, Copernicus Books, https://doi.org/10.1007/978-1-0716-2776-1_2

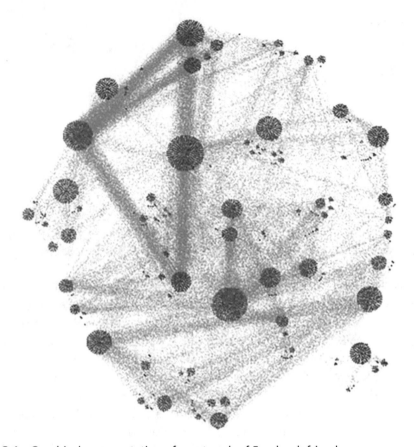

Fig. 2.1 Graphical representation of a network of Facebook friends

friends or new colleagues). As a result, there are also mutual connections between the various clusters, but there are far fewer connections between clusters than within clusters.

The friendship paradox: why do my friends have more friends than I have?

We all do our best to take care of our network. We make contacts, send messages, follow and 'like'. Yet most of our friends seem to have more friends than we do. That's not just a feeling, it's backed up by facts. No less than 93% of all Facebook users are in that situation! In 2012, an extensive study of all Facebook users – and there were around 721 million of them back then, with around 69 billion friendships between them – showed that a Facebook user has an average of 190 friends. At the same time, the same average Facebook user sees that his friends have an average of 635 friends! One would feel terrible for less.

(continued)

Yet, this is not at all as strange as it seems at first sight. Let's look at this phenomenon in detail for a clique of 4 friends: Alice, Bob, Carol and Dave, who are connected, as in Fig. 2.2, and the loner Ernie who has no friends. We all read the number of friends on the figure and find the average number of friends in this network: (2 + 1 + 2 + 3 + 0)/5 = 1.6. Now look at the average number of friends everyone sees: Alice is friends with Carol and Dave. So she has two friends, who themselves have 2 and 3 friends, respectively. An's friends have an average of 2.5 friends, while An only has 2 friends of her own. For Bob it is even worse. Bob only has 1 friend (Dave) and that friend himself has 3 friends. For Carol, the same applies as for Alice. So her friends also have on average 2.5 friends, while she only has 2 friends of her own. In our example, only Dave's friends have fewer friends on average than Dave himself, namely (2 + 1 + 2)/3 = 5/3 = 1.66.

The crux of the matter is that people with many friends themselves are more often on someone else's list of friends. Just look at the example above: Dave

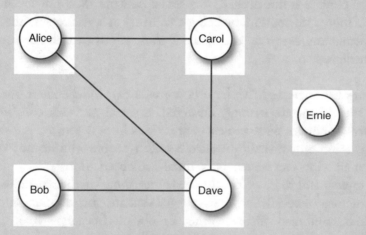

Fig. 2.2   Graphical representation of a small network of five friends

(who has three friends) was mentioned exactly three times, namely by his three friends. And Bob (with his one friend) was mentioned only once, exactly by that one friend. To make it completely clear: take the extreme case of someone without friends, for example Ernie. The fact that Ernie has zero friends did count towards the average number of friends of the group of five. But when calculating the average number of friends of friends, Ernie is ignored because he does not appear on anyone's list of friends. So, people who have many friends of their own are counted more often (and hence are more important for the average) when we count friends of friends.

The world is small and strongly connected   Despite the limited number of connections, it often doesn't take long to get a message from any person to any other person. This is something we have known for a long time. In 1967, long before social media existed, the American psychologist Stanley Milgram

(1933–1984) conducted an experiment. He gave a letter to a number of randomly selected people in a number of remote places in the United States (such as the city of Omaha in Nebraska) and asked them to deliver that letter to a specific recipient in Sharon, MA, a suburb of Boston. However, they were only allowed to pass the letter on to people they knew personally. Milgram then counted the number of steps needed for the letter to arrive at its destination. That number turned out to be surprisingly small: somewhere in the neighbourhood of six. (Of course, there were also letters that never arrived. Those were not counted.) If we would repeat this experiment today (with an e-mail instead of a letter), it seems very likely that fewer steps would now be needed. First, because there are now fewer geographical limitations between friends: in Milgram's experiment, physical proximity was required, while physical contact is not needed on a social network. Second, social networks provide more information: by being able to see at a glance who the friends of our friends are, everyone can make a more motivated choice for the next intermediate step.

Looking for structure We have now made two independent observations. First, friendships are strongly clustered: a social network consists of large numbers of cliques with weaker connections in between. At the same time, there is often only a small distance between people who do not know each other at all. How can we reconcile these two observations?

To understand that, we clearly need some more mathematical insight into the structure of social networks. To get that insight, different mechanisms need to be combined with each other. An obvious first step is to quantify what types of connections between nodes typically occur in a social network. Then you look at real data from a social network, and find out, for example, how many friends someone has on average, what the fraction of people with a specific number of friends is, or how often people have friends in common.

The next step is to create our own artificial networks. We want those to have a simple structure, containing just those properties that we suspect to be important to explain the real-life observations we made. We will now look at two extreme artificial networks. Imagine first of all that the friendships would be random, as shown on the right in Fig. 2.3. In that figure, each person is on average friends with four others, but there is no connection between the friendships. In such a network we can very well explain why average paths are short. If you have 100 friends and they each have 100 friends, then you can reach 10,000 people in two steps. In three steps, you can reach a million people. However, that network does not have the structure of a social network. On a real social network, many of your friends' friends already belong

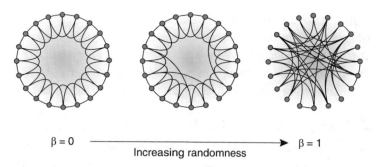

$\beta = 0$ ⟶ $\beta = 1$
Increasing randomness

Fig. 2.3 Different types of networking between friends. On the left everyone knows his two left neighbours and his two right neighbours. On the right, on average, everyone knows four other people at random. In the middle, everyone knows mainly his neighbours and occasionally a random other person

to your circle of friends. So in two steps, you reach far fewer than the theoretical maximum of 10,000.

In the other extreme case, there are only connections between people who are in each other's 'neighbourhood'. That neighbourhood can be something physical, such as houses in the street, or something else that brings people together, like the school. On the left in Fig. 2.3, everyone is friends with the two people on their left and with the two people on the right – a kind of mathematical version of the 'us knows us' principle. In such a network, there are a lot of interconnections between your friends' friends. When two of your friends are also friends, a 'triangular relationship' arises in the network. Because the word 'triangular relationship' might hint at a 'love triangle', we will use the more 'serious' term 'triad'. The number of triads in a social network is a very important parameter when investigating the characteristics of social networks.

Small World Networks Unfortunately, a network such as the one on the left in Fig. 2.3 does not have short paths between any two people. We must therefore look for a way to create artificial networks that at the same time count a large number of triads and have short paths between any two people. A major breakthrough came when the Americans Watts and Strogatz proposed in 1998 to take a combination of the two extreme cases: a number of connections to neighbours that represent the 'us knows us' principle and a number of random connections that describe distant acquaintances. To realize this, Watts and Strogatz started from a completely regular network like the one on the left and replaced a certain fraction of the connections with random ones. The networks they created in this way can be fully described with only two numbers: the number of neighbours to which everyone is initially connected, and the fraction of random connections. The two previous extreme cases are now special cases of the general Watts–Strogatz network, specifically the cases where the fraction of random connections is zero (left) or one (right). Networks as created by Watts and Strogatz are called small world networks.

Analysis of small world networks Of course, real social networks don't look like the ones suggested by Watts and Strogatz. Indeed, the two mathematicians were not interested in representing the structure of a social network as accurately as possible. Instead, they wanted to build as simple a network as possible that contains the essence of a social network – and *only* that essence – and for which they could study some properties theoretically. In this way, they proved that you can typically move from one node to another node in such a network with a limited number of intermediate steps. To be specific, the number of necessary intermediate steps increases with the logarithm of the number of people in the network, provided the network is large enough. The logarithm is a very slowly increasing function: if the network size doubles, the number of necessary steps only increases by one! This result implies that the small world principle can be explained by using only a limited number of distant acquaintances. Small world networks are often used as a simple model to understand certain phenomena on social networks. For example, they demonstrate the importance of a limited number of highly connected individuals for information dissemination. And when we allow friendship connections to evolve over time, they can be used to study how small clusters of radicalized individuals emerge. This follows, for example, from adding a simple rule that states that people will have a small probability of breaking a friendship in case of a disagreement and/or to form a new friendship with a like-minded person.

My enemy's enemy is my friend We can go a step further and give a sign to the connection between two people: positive in case of friendship, negative in case of enmity. When two friends of the same person are also friends, we speak of a balanced triad. Another triad, in which the two enemies of the same person are friends, is also balanced. Unbalanced is a situation where a person is friends with two persons who are mutual enemies. (Think, for example, of a common friend of a couple that went through a venomous divorce, as shown in Fig. 2.4). Now it appears (in real life) that individuals in an unbalanced triad are motivated to adjust their attitudes. (In the example, this can be done by choosing one side, reconciling both parties, or severing all contact with both). In the mathematical model, the connections in the network then change at a certain point in time. The American Harary demonstrated in

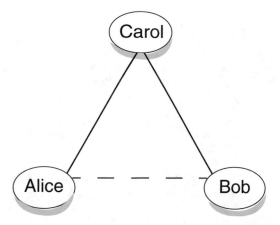

Fig. 2.4 Graphical representation of a network of three people. Carol is friends with Alice and Bob, who have a hostile relationship with each other

1953 that a fully connected network can only be balanced if everyone is friends, or if two camps exist, with only friendships within their own camp and only enmities with the other camp. Only then will each triad in the network be balanced. Therefore, a situation with three camps can never be stable.

To illustrate the natural tendency towards a balanced network, we can look at the evolution of the alliances in the run-up to the First World War, as shown in Fig. 2.5. We see a gradual evolution towards a situation with two camps. As soon as this situation is reached, the alliances are stable. (The result of course was that the stability of the world deteriorated strongly. Because there was no party left that could mediate, we automatically ended up in a world war.).

Detecting communities in Belgium Belgium is a small country with 11 million inhabitants and a reasonably complicated structure: the north of the country is Dutch-speaking, the south French-speaking and Brussels, the capital, is bilingual. There is also a community of 70,000 German-speaking Belgian in the east of the country. Since Belgium is a relatively recent country (founded in 1830), its internal structure is still evolving. The main point in these discussions is to find a good equilibrium between centralized and decentralized authority. So, it is natural to ask the question what would be a coherent division of the country into smaller regions. Currently, there are 10 provinces (indicated by black lines on the map), and the question is: can we use a network analysis to determine whether these provinces are reasonable? The Belgian Université Catholique de Louvain processed 200 million telephone calls between the 2.6 million subscribers of a major mobile telecom operator. Based on the frequency and duration of these calls, a certain impor-

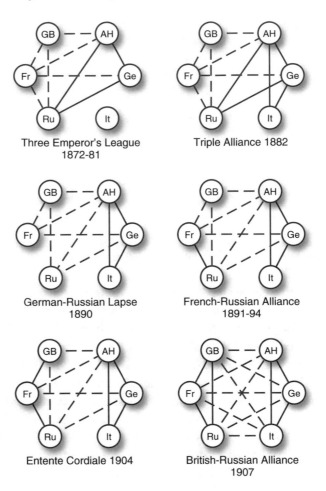

Fig. 2.5 Evolution of military alliances in the run-up to World War I. Black lines represent bonds of friendship. Red lines are enmities. Eventually the countries were divided into two strictly separated camps. The network is balanced. The war can begin. GB Great Britain, AH Austria-Hungary, Fr France, Germany, Ru Russia, It Italy

tance was attached to the intensity of the connection between each two people. Subsequently, the cliques (or clusters) in this huge network were detected. The astonishing result is shown in Fig. 2.6: the clusters correspond perfectly to geographically contiguous municipalities. Even the provincial boundaries are found well (but not perfectly), while this was not imposed at all by the method, even if the algorithm indicates that there would rather be 17 provinces than the 10 we know today. So, the subdivision into provinces is not that unreasonable.

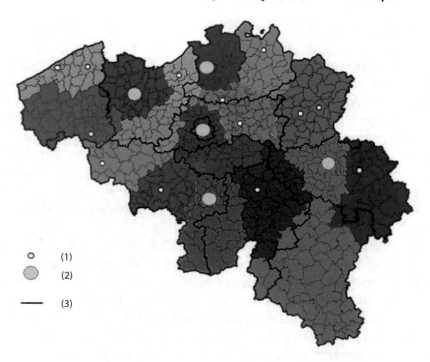

(1)

(2)

(3)

Fig. 2.6 Breakdown of Belgium into 17 sub-areas based on an analysis of the frequency and duration of telephone calls made by 2.6 million subscribers of a mobile telecom operator. Thin lines indicate municipal boundaries; the black lines are the provinces. The areas correspond more or less to parts of the current provinces, with some notable exceptions. Vincent Blondel, Gautier Krings and Isabelle Thomas, "Regions and borders of mobile telephony in Belgium and in the Brussels metropolitan zone", Brussels Studies, 2010. https://doi.org/10.4000/brussels.806

A thorough understanding of the characteristics of social networks has many concrete applications: it enables one to estimate the effect of viral marketing campaigns, to find out which people belong to a common target group or to identify important individuals. However, the scope of small world networks is much broader than Facebook and Twitter. This type of idea is quite universal, and also proves its usefulness in other applications. Amazon, for example, uses similar algorithms to network their book catalogue so that they can efficiently recommend a number of interesting other books to you with each purchase. Airlines use the structure of small world networks to select their airports and hubs. And there are even indications that the neurons in our brains make connections that are very similar to small world networks. So new theoretical insights into the properties of small world networks can suddenly lead to breakthroughs in many domains.

Teapots, Ponytails, and an Unexpected Reward

A leaking teapot can turn even the most beautifully set table into a wet, dirty mess in seconds. Often you have the feeling that you just can't avoid it. Whether you pour quickly or slowly, stop the jet gently or abruptly, there will always be a puddle next to the cup. Why is that? Most scientists will turn their eyes away when they hear that question, mumble something about surface tension and then quickly get out of the way. Logical, because they don't really know exactly. That surface tension certainly doesn't play a role can be verified by doing a small experiment in which you switch off that effect. Scientists have really tried that, and they have noticed that a teapot also drips when you pour (coloured) tea under water.

The American mathematician Joe Keller (1923) proved in the 1960s that the dripping of the teapot can be mathematically fully explained by calculating the trajectory of a ray of tea from a teapot in great detail. With this result, the teapot – after 30 years of follow-up research in the margins – could finally be perfected in the 1990s by applying a thin black layer on the end of the spout. It earned Keller an Ig-Nobel prize in 1999, a prize for serious scientific research that first makes you laugh, but then also makes you think.

Joe Keller is exceptional. He is the only double winner of an Ig-Nobel prize, but also one of the most important contemporary researchers in applied mathematics. He has made contributions in numerous fields, including pioneering work in the understanding of wave propagation. His work, for example, made it possible to calculate the effect of an underwater explosion of an atomic bomb, or to design submarines so that they became invisible to radars. In addition to his two Ig-Nobels, he received many serious awards, including the *National Medal of Science* from the hands of the U.S. president.

Joe Keller received his second Ig-Nobel prize in 2012. He was already over 80 at the time, but still a passionate runner. The work that led to his second prize was created while he was jogging in the morning. As he watched the joggers in front of him, he asked himself the following question: if the runner's head goes up and down, why does the ponytail go from left to right? The key to the answer is to calculate the movement of a bendable wire that is shaken at one end. The calculations show that the effect is most pronounced for a ponytail of about 25 cm long. By the way, the mathematics he used to obtain this result also had more serious applications. It formed the basis for the scientific research that earned Wolfgang Paul and Hans Dehmelt the Nobel Prize in Physics in 1989.

Chapter 3
Are the Polls Right?

Since they were introduced in the 1920s in the United States, the popularity of opinion polls has only increased. Nowadays, we can't even escape them at times when there are no elections at all. But how reliable are these polls? And what conclusions can we draw for the distribution of seats in a parliament or the US Electoral College?

What's the problem? Elections are a logistical nightmare: every individual entitled to vote has to cast their vote on one particular day and all those votes have to be counted the same day. This involves a lot of manpower, voluntarily or otherwise, and the necessary infrastructure is considerable. Elections are therefore – in principle – only held once every few years. In between, polling stations occasionally ask a limited number of people to make their voting intentions known, to get an indication of evolutions in the whole population. Because opinion polls are quite expensive, the sample is usually limited to about 1000 people. The result of an opinion poll is therefore only an estimate of the voting intentions, with limited accuracy. In addition, the final result of an election is not the voting percentages, but a distribution of seats in a parliament or an electoral college. In some countries, such as the US, the UK and France, a winner-takes-all principle is used: the territory is divided into small regions that each elect one or a few representatives. The person or party with the most votes then gets the representatives. Other countries use a proportionality system: the available seats in each region are then distributed according to the percentage of votes. To minimise the number of 'useless' votes, complicated rounding mechanisms are used to map the voting percentages into a seat distribution. This, in turn, implies that a few votes can overturn the seat distribution. All these considerations lead to a number of important questions: what is the value of an opinion poll? How can the margin of error be kept as small as possible? And how exactly can the seats be fairly distributed?

G. Samaey, J. P. L. Vandewalle, *The Invisible Power of Mathematics*, Copernicus Books, https://doi.org/10.1007/978-1-0716-2776-1_3

The Belgian parliamentary elections of 2014 We start from the 2014 election results for the Belgian Chamber of Representatives. Belgium has a proportionality system, and every party that collects more than 5% of the popular vote in a particular province gets at least a seat. Moreover, Belgium is a bilingual country, with Dutch as the language in Flanders, and French as the language in Wallonia. There is also a German-speaking region in the East, but let's ignore that here. The structure of the country is much too complicated to describe in this chapter, and would be the topic of a book in its own right. Let's suffice here by saying there are many political parties in Belgium, which makes it an interesting example for this chapter. To keep things a bit manageable, we only look at the votes for the seven largest parties in Flanders[1]. Let's start with a simple test based on a small computer program. The results can be found in Table 3.1. With our computer program we now conduct two fictitious opinion polls. In each of these virtual polls, the computer mimics 1000 virtual voters that each votes according to the real election result. So, for each of those virtual votes there is, for instance, a 32.5% chance that it goes to N-VA and an 8.5% chance that it goes to Green. The results of these virtual polls are also shown in the table.

You immediately notice some striking differences between the opinion polls and the real results: both polls give a big win for NV-A, but still the predicted results differ by more than 3 percentage points. In the first poll, there is a large gap between Open VLD and sp.a. This gap is not present in the second poll nor in the election results. In addition, the predicted results for Open VLD, sp.a. and Vlaams Belang in the second poll are virtually the same as the election results. The second poll therefore seems to be clearly better than the first. Still, because of the setup of the experiment, they are both

Table 3.1 Results of the elections for the Belgian Chamber of Representatives in 2014 for the seven largest Flemish parties, together with the results of two virtual opinion polls in which 1000 random votes were generated. Each of the votes in the opinion polls is allocated to a party with a chance that perfectly matches the election result

	Election results	Virtual poll 1	Virtual poll 2
N-VA	32.5	31.1	34.5
CD&V	18.6	19.1	17.2
Open VLD	15.7	17.1	15.7
sp.a.	14.2	13.2	14.3
Green	8.5	7.6	7.9
Vlaams Belang	5.9	6.6	5.9
PVDA	2.8	3.6	3.2

---

[1] For those interested: the traditional parties are the Christian-democrats (CD&V), the liberal democrats (Open VLD), and the socialists (sp.a). Additionally, there are two nationalist parties that strive for independence of Flanders: the conservative N-VA party and the anti-immigrant party Vlaams Belang. The ecologists are called Groen (Green in English), and the communists PvdA (or Labour Party in English). There are more parties than this, but let's stop here.

equally random, as we know. So these detailed differences are interpretations that simply are not realistic. In reality, the presented conclusions are routinely made with this type of opinion poll. Even more problematic, if opinion poll 1 and opinion poll 2 were carried out in consecutive months, one could conclude from Table 3.1 that certain parties are 'improving' in the polls. However, both polls were taken at random from the same population and, moreover, the voting behaviour of that population corresponds perfectly to the election results. The conclusion is clear: all differences between these two polls – and between those polls and the election results – are due to coincidence in the selection of the virtual voters in the sample.

Understanding opinion polls with statistics

To give some insight in the estimated statistical error of an opinion poll, we start again from the election results for the Belgian Chamber of Representatives in Table 3.1. Instead of conducting two virtual opinion polls with our computer program, we now do 10,000 of these polls, each with 1000 randomly generated virtual voters. Once again, the voters' voting behaviour is entirely in line with the election results.

Now we take the next step: we calculate the probability that a virtual opinion poll will produce a specific result for N-VA. Figure 3.1 shows the results as a histogram. On the horizontal axis, we put the possible voting percentages of N-VA in the virtual poll, divided into small intervals with a width of 0.2%. On the vertical axis, we put the number of opinion polls (of our 10,000 conducted) in which we actually measured that specific voting percentage.

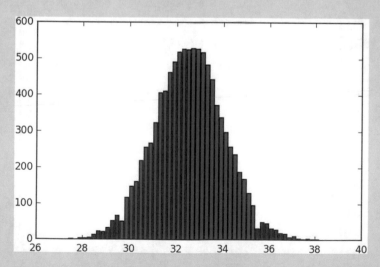

Fig. 3.1 Distribution of the results of an opinion poll with 1000 virtual votes based on the exact election result. The 5% opinion polls that deviate more than 2.9 percentage points from the real result are shown in red

(continued)

Two things stand out immediately: the figure is symmetrical around the actual election result of 32.5%: an overestimation and an underestimation are equally likely. Moreover, there is a reasonable probability of an error of a few percent in the poll. In 95% of virtual opinion polls, the measured value appears to have a deviation from the election result of less than 2.9 percentage points. The 5% virtual polls with a larger deviation are shown in red. We then say that an opinion poll with 1000 respondents has a margin of error of 2.9 percentage points for this result, with a reliability of 95%.

Fortunately, we can also determine the margin of error without having to carry out thousands of virtual opinion polls because there is a mathematical expression for it. We look at a sample of N virtual voters and an election result, p noted as a fraction. (The result of, for example, Open VLD then becomes p = 0.157.) The 95% confidence interval for this p is then given as $[p - 1,96\sigma; p + 1,96\sigma]$, containing $\sigma = \sqrt{p(1-p)/N}$.

We can make two more observations. First, the confidence interval decreases as the sample size (the number of voters in a single opinion poll) increases. This is of course logical, as we can see in Fig. 3.2. There, the previous experiment was repeated with virtual opinion polls with 10,000 virtual voters instead of 1000. Second – and perhaps more surprisingly – the size of the total population plays no role in determining a good sample size. This is because the formula is derived from an infinitely large population. This is a reasonable approximation as soon as the population is several thousand times larger than the sample. For example, the formula can be used to show that, in an election between two candidates, a sample of 4001 voters will correctly predict the winner in 99% of the polls if the victory is not close. By close, we mean that the winner gets between 50% and 52% of the votes.

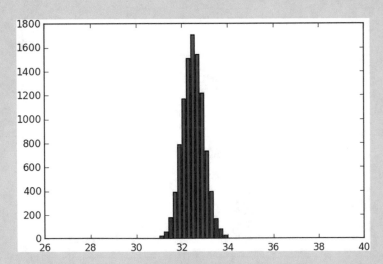

Fig. 3.2 Distribution of the results of an opinion poll with 10,000 virtual votes based on the exact election result. The 5% of opinion polls that differ by more than 2.9 percentage points from the real result are shown in red

How should we read polls? That polls do not say everything about an election result is obvious, but that they do not teach us anything is also not correct. To interpret the results of an opinion poll properly, we need a little more insight into the accuracy of the estimated results. Often, the accuracy of an opinion poll is not mentioned in the media. If something *is* mentioned in this respect with poll results, it is often claimed that there is 'a 3 percentage point margin on the results for a 95% significance level'. This is a piece of specialist jargon which means that the result of an opinion poll in 95% of the polls carried out will not differ by more than 3 percentage points from the actual election result. As is illustrated in the frame, we can demonstrate that this standard is met as soon as the sample contains 1000 participants, regardless of the size of the population. So, be warned: in 5% of the polls, the election result will deviate from the poll by more than 3 percentage points. This is unavoidable statistics. A large deviation does not automatically mean that the poll was badly conducted!

It immediately follows from these findings that displaying the first decimal of polling results is utter nonsense. The first decimal represents the poll result up to tenths of a percent and we already know that we have to take a margin of 3 percentage points on the result obtained. Additional digits therefore do not contain any useful information. Some opinion polls do explicitly include the 3% margin of error in the reported percentages and then write $32.5 \pm 3\%$ for N-VA, for example. This is mathematically correct, but still puts a lot of emphasis on the measurement. It is best to give a range, as is done in Table 3.2 below:

Table 3.2 Results of the elections for the Belgian Chamber of Representatives in 2014 for the seven largest Flemish parties, together with a representation of the results of two virtual opinion polls with 1000 random votes that correctly reflect the uncertainty of the opinion poll

	Election results	Opinion poll 1	Opinion poll 2
NV-A	32.5	28.2–34.0	31.6–37.4
CD&V	18.6	16.7–21.5	14.7–19.6
Open VLD	15.7	14.8–19.4	13.4–18.0
sp.a.	14.2	11.0–15.4	12.1–16.5
Green	8.5	5.9–9.3	6.2–9.6
Vlaams Belang	5.9	5.1–8.1	4.4–7.4
PVDA	2.8	2.5–4.6	2.2–4.2

Both polls are clearly consistent with the election results, but you can also see that the nuance in the last table does not easily translate into sensational newspaper headlines.

The American presidential elections of 1948 The computer experiments above were carried out with a perfectly homogeneous population, with each person voting in the same way. In reality, this is not the case. Voting behaviour is related to many factors such as work, level of education or gender. We must then ask ourselves whether the sample used in the poll reflects the diversity of the population. If it does not, we call the sample unrepresentative. As a reminder, the problems with the limited accuracy described above are inherent to an opinion poll, even if the sample is perfectly representative. Errors in the composition of the sample can only lead to additional problems, which we call bias. An example of such a selection bias is the American presidential elections of 1948, which was between the Democrat Harry Truman and the Republican Thomas Dewey. The polls were very clear: Dewey would win, and by a wide margin. (Table 3.2 shows the estimates of three different polls.) Due to the size of the samples, the statistical error was very small. Everyone believed these polls so much that certain newspapers, including the Chicago Tribune, put Dewey's victory on the front page even before the votes had been counted. So, the bewilderment was great when the results rolled in and Truman turned out to be the winner.

	Election results	Gallup	Roper	Crossley
Harry Truman. (Democrat)	**49.6%**	44.5%	38%	45%
Thomas Dewey (Republican)	**45.1%**	49.5%	53%	50%

Now, what went wrong? It turned out that the composition of the sample was not representative for the entire population. For example, the polls were partly conducted by telephone and in 1948 the telephone was not yet widely distributed, certainly not among the worker class. The polls therefore only predicted that Dewey would win with telephone users, not with the entire population! An important difference.

Selection of representative samples The above example points to a general problem with opinion polls: it is important that the 1000 members in the sample are representative of the diversity of the population. Suppose, for example, that 52% of the population are women, that 38% of the population is aged between 25 and 44, or that 32% belong to a certain socio-economic class. We then want to see these numbers reflected in our sample. When the participants have been randomly selected, this is solved by asking additional questions to determine in which category each participant belongs. An imbalance in the sample is then dealt with by weighing the results appropriately. Suppose, for example, that our sample contains only 48% women, whereas we know that 52% of the population consists of women, then a vote of a woman in our poll will be given a slightly higher weighting to compensate for the underrepresentation of women in our sample. It is of course important to choose the categories in such a way that they are relevant to voting behaviour. The factors that are usually considered important are gender, age, level of education, place of residence and occupation. Based on the place of residence, sometimes the last election result is also used as a reference for the re-weighting!

What else could go wrong? Selection bias is not the only problem. Another tricky issue is the treatment of undecided voters. In 1948, they were simply removed from the sample. This is clearly not a perfect option because it means that the votes of this group are distributed in the same way as those of the decided voters. Another problem is that America does not have compulsory voting. So, there is no guarantee that someone who expresses a preference for a candidate on the phone will actually vote. And the fact that people lie about their preference is, of course, completely unsolvable. That last problem is very difficult to take into account using a mathematical approach because it is not clear in which direction the bias will occur. Some voters are bolder during an opinion poll than in the polling booth, while others are just ashamed to make their voting public. (In Flanders, for example, the anti-immigrant party Vlaams Blok (later condemned for racism and rebranded as Vlaams Belang) was systematically underestimated in the polls of the 1990s).

The paradoxes of the distribution of seats

Even with the most accurate prediction of voting rates, it remains extremely difficult to make a precise estimate of the resulting distribution of seats in parliament before all results are in. Election results have to be rounded because a seat always goes in its entirety to a particular party. This rounding is not as simple as it seems. The simplest working method was devised by Alexander Hamilton, one of the Founding Fathers of the United States. His idea was to calculate the number of seats first without rounding (i.e., as a decimal). That result is called the 'fair share'. Then these 'fair parts' are all rounded down. As a consequence, there will be a number of seats left. These go to the parties that had the largest remainder at rounding off (one seat per party). In this way, the 'fair share' for each party is rounded up or down.

(continued)

The Alabama Paradox Hamilton's system seems perfectly reasonable and fair, but can lead to strange paradoxes. For example, parties that obtain an equal election result as the previous time may actually lose a seat when the parliament as a whole gains a seat. This is called the Alabama paradox because the state of Alabama was the first to suffer from it. In the US Congress, states are proportionally represented according to their people. When the US Congress expanded in 1882, it cost the state of Alabama a seat. A simple example illustrates the problem. The table shows three states (A, B and C) with their population numbers. When adding a seat in Congress, state C loses its second seat.

State	Population	10 seats		11 seats	
		Fair share	Seats	Fair share	Seats
A	6000	4.286	4	4.714	5
B	6000	4.286	4	4.714	5
C	2000	1.429	2	1.571	1

This seems a rather rare situation, yet the enlargement of a parliament occurs regularly. Just think of the growth of the European Union. Hamilton's method also contains other paradoxes: for example, the 'population paradox' can, in certain circumstances, lead to a party losing seats, even if it greatly improves its voting percentage.

The impossibilty theorem of Balinski and Young In 1982, the mathematicians Balinski and Young proved an important theorem: there can be no electoral system that works with rounding and at the same time is free of the Alabama paradox and the 'population paradox'. Of course, we can also speak of a paradox if a party obtains a number of seats that does not correspond to a rounding off of its fair share. Then the theorem actually says: every method of allocating seats contains paradoxes. That's disappointing, of course, but if this is the case, it is best to be aware of the unavoidability.

Quotient tracing methods In general, the Alabama paradox and the population paradox generate the most resistance from public opinion. To counter them, methods have been developed that use two basic principles:

1.   the more votes, the more seats;
2.   the number of seats allocated is 'near' the fair share.

It automatically follows from the previous discussion that we will have to apply some flexibility to the second principle if we want to avoid strange paradoxes. Quotient drawing methods are a general way to do this. In such methods, the seats are distributed one by one by dividing the number of votes by a series of divisors. We construct the quotient series for each party by dividing the number of votes by the corresponding divisor in the divisor series, as shown in the table below. Then we put the quotients in order from large to small. The seats are allocated one by one to the largest remaining quotient, and this until all seats are used up. In the d'Hondt method, which is used in the Belgian parliamentary elections, the series of divisors is 1, 2, 3, etc. For the Imperiali method, used in municipal elections, the series is 2, 3, 4, etc.

(continued)

In general, this results in fairly proportional distributions. The specific choice of divisor series does have an impact on rounding and can lead to strange results in exotic cases. An example: In the table, 10 seats are divided between parties A, B and C according to their number of votes. We use the d'Hondt method.

	Party A		Party B		Party C	
Votes	60,000		25,000		15,000	
Percentage	60		25		15	
Fair share	6		2,5		1,5	
Divisor	Quotient	Seat	Quotient	Seat	Quotient	Seat
1	60,000	1	25,000	3	15,000	6
2	30,000	2	12,500	7	7500	
3	20,000	4	8333		5000	
4	15,000	5	6250		3750	
5	12,000	8	5000		3000	
6	10,000	9	4166		2500	
7	8571	10	3571		2142,	
8	7500		3125		1875	

We notice that the number of seats for both party B and party C is rounded down and that party A gets seven seats, although the fair share for party A did not have to be rounded at all!

Advantages for large parties In general, the most commonly used methods result in an advantage for large parties. This happens more easily when the first divisor is larger (as with the Imperiali method) or when the opposition is divided into many small parties. It then regularly happens that the largest party can get the absolute majority of the seats without having the absolute majority of votes. In some countries, such as Greece, even a fixed number of seats are allocated to the largest party as a bonus, so that a narrow election victory can still produce a comfortable majority.

Opinion polls are an important means of informing policymakers and the public about voting intentions. Moreover, there are strong indications that opinion polls also strongly influence the voting behaviour and the behaviour of politicians. Doubtful voters can get carried away, for example, to 'follow the masses' and decide to also vote for the party that is growing in the polls. Strategic voters may decide to vote for parties that seem to end up close to the electoral threshold (or to do the opposite if they fear their vote would be lost). Politicians can adjust their message based on the results. However, because statistical errors in the results of opinion polls are inevitable and distortions are difficult to detect, the results always contain uncertainty and bias. A little (mathematical) insight into the nature of that uncertainty is therefore absolutely necessary to interpret the polls correctly.

Chapter 4
Cochlear Implants Help the Deaf Hear Again

The ear is an ever-active organ, constantly picking up and processing sound signals. It signals danger, allows us to talk to each other and ensures that we can enjoy music. But how does hearing work exactly? And how can a hearing aid take over this function?

What's the problem? Sound waves are vibrations of air, caused by pressure differences. These vibrations can occur at different frequencies. Each frequency corresponds to a specific pitch. Most sounds consist of a whole range of frequencies. Our ear initially absorbs sound waves in their entirety via the eardrum. These waves are then transmitted mechanically through the ossicles to the cochlea in our inner ear. The cochlea is a coiled tube that contains vibrating hairs (called cilia) that vibrate at specific frequencies in the sound signals. When that happens, an electrical pulse is transmitted to the corresponding auditory nerve fibres. These electrical stimuli are then processed by the brain into what we experience as 'hearing' sound. For some people, however, this mechanism no longer works as it should. Can we replace the functioning of this complex organ with a piece of electronics? And can these electronics work so fast that the wearer no longer notices that it is there? (Fig. 4.1)

Meningitis and hearing problems Hearing problems can arise in many ways. The sensitivity of the entire hearing organ can be affected by excessive exposure to loud music or noise. The auditory nerve can be damaged, for example, by brain surgery. And the cilia can be destroyed by disease. Children who contract meningitis, for example, receive increased pressure in the brain that can destroy the cilia in their inner ear. In all these cases, hearing loss, and in the worst case, deafness, is the result. As long as the auditory nerves are still working, a technological solution should be possible in principle: in that case, the auditory nerves can still transmit electrical pulses to the brain. We just have to make sure that the right electrical pulses reach that auditory nerve. If

G. Samaey, J. P. L. Vandewalle, *The Invisible Power of Mathematics*, Copernicus Books, https://doi.org/10.1007/978-1-0716-2776-1_4

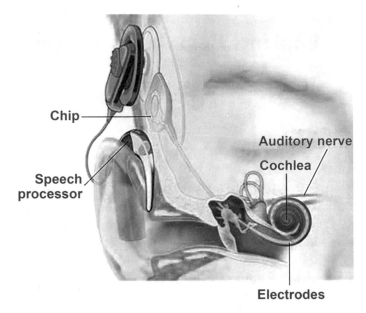

Chip

Auditory nerve

Cochlea

Speech processor

Electrodes

Fig. 4.1 Sketch of an ear with cochlear implant

damage to the vibrating hairs is the problem, the idea is thus to make an implant that takes over the function of the vibrating hairs and stimulates the auditory nerves electrically.

How do the vibrating hairs work? In a normally functioning ear, the vibrating hairs or cilia behave like the ears of corn on a field, which are rocked by the wind. Each vibrating hair has a natural tendency to vibrate at a frequency determined by its shape and length. Now the incoming sound is a signal that consists of a lot of different frequencies. A vibrating hair then behaves like a kind of sieve: it selects those parts of the incoming sound that correspond to its own natural frequency and transmits an electrical pulse to the corresponding auditory nerve fibre. The size of that electric pulse is then proportional to the strength of that frequency in the incoming sound. The shape and length of the vibrating hairs varies in the cochlea. For all audible frequencies in the incoming sound, a specific group of vibrating hairs resonates with the corresponding natural frequency. The brain then has the opposite task: it receives from the auditory nerves the decomposition of a sound signal in its frequency components, and has to let us experience the entire sound signal.

Replace the cilia by a cochlear implant To replace the vibrating hairs when they no longer work, a device has been developed that is implanted in the ear, the cochlear implant. This implant converts the ambient noise into electrical signals that can be received and processed by the auditory nerves, thereby skipping the inner ear. It consists of an external part, which is shaped like a classic hearing aid and is worn behind the auricle, and an internal part that is placed in the cochlea. The external part contains a microphone that captures the sound and a speech processor that calculates how the captured sound should be presented to the internal part. The speech processor transmits the result of its calculations wirelessly to a microchip in the inner part. That microchip in turn generates electrical pulses in a very thin cable full of electrodes that are directly connected to the auditory nerves. In this way, the auditory nerves receive an electrical stimulus, which they transmit to the brain. In the brain, the entirety of these signals is recognized as sound. To make this system work properly, two mathematical problems need to be solved. The first problem is of a theoretical nature: how can we decompose a sound signal into individual components with specific frequencies? The second problem is more practical: how can we calculate this decomposition so quickly that a patient does not experience delay while listening?

Sound signals consist of basic waves of specific frequencies We can make a mathematical representation of a sound wave by writing down a function that represents the evolution of the air pressure in the ear as a function of time t. A vibration at a single frequency is described by a sine or cosine function, A sin (ωt) or A cos (ωt). We call A the amplitude and ω the frequency of the vibration. The sine and the cosine represent the same frequency, with a phase shift of 90°. If we now receive two sound signals at the same time with a different frequency, we will hear it as one sound signal containing those two frequencies. This implies that we may simply add two sound signals. In general, an arbitrary sound signal can always be considered as a sum of a (possibly infinite number of) sines and cosines with an appropriate frequency and amplitude. The French mathematician Jean-Baptiste Joseph Fourier (1768–1830) realized this and also drew up the formulas to decompose any random signal again in those basic waves. The dissolution of an arbitrary signal in its basic vibrations (or Fourier modes) is therefore called the Fourier transformation. It is nice to note that Fourier invented his method long before cochlear implants. For him, this observation was a necessary step in his research into the diffusion of heat (Fig. 4.2).

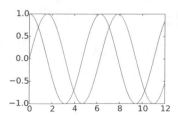

Fig. 4.2 Sine (red) and cosine (blue) with amplitude and frequency 1

The idea of Fourier

The Fourier transformation is a mathematical decomposition of a signal (such as the sound reaching the ear) into the various frequency components it contains. Let's look at the case where the signal is periodic in time with period 2π. (That period is the period of the sine and the cosine with frequency 1). The Fourier transformation writes this signal as an infinite sum of sines with *multiples of 1* as frequencies. In this way, the signal is 'decomposed' into its basic components per frequency (which we call the Fourier modes). Fourier proved that any signal can be decomposed uniquely that way. Moreover, he derived the formulas to actually calculate this decomposition.

We illustrate the Fourier decomposition for a signal that looks like a block function, the blue line on the drawings on the left in Fig. 4.3. On the right in Fig. 4.3, you can see the first six Fourier modes. You will notice that the frequency of these modes is increasing: there are more and more vibrations per unit of time. Moreover, their amplitude is getting smaller and smaller: the contribution of higher frequencies to the complete signal is getting smaller and smaller. In the other figures on the left, you can look at the approximation of the block function with three Fourier modes (centre) and with six Fourier modes (bottom). You will observe that the approximation improves by taking more Fourier modes. From a certain frequency onwards, the contribution of the following Fourier modes becomes negligible.

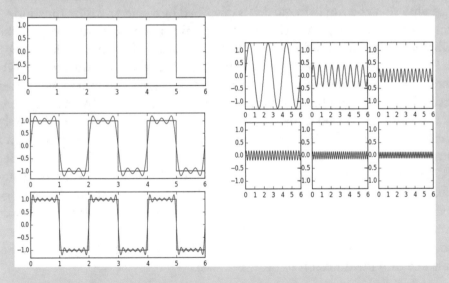

Fig. 4.3 Sketch of the main aspects of the Fourier decomposition. Top left: block signal in blue. Right: the components of this block signal according to the first six basic frequencies. The faster the vibrations, the higher the frequency. The larger the amplitude, the more the wave contributes to the block signal. Left middle and bottom: again the original block signal, together with an approximation of this signal with the first 3 (middle) and 6 (bottom) base waves

From theory to practice The speech processor must calculate the Fourier transformation for the incoming sound, in order to stimulate each of the auditory nerve fibres with an appropriate intensity. Although it is mathematically clear *what* needs to be calculated, it is not at all clear *how to* do it as quickly as possible. It goes without saying that it has to be done very quickly: the slightest delay on those transmitted signals is very annoying for the wearer of the implant. On the way to this goal, however, we still have to overcome a number of practical problems. First of all, there is the observation that according to the Fourier transformation, the incoming sound is split up into an *infinite number of* modes. To calculate them all, of course, an infinite amount of computation is needed – something no computer can do in a finite amount of time. Fortunately, people only hear frequencies up to 20,000 Hertz. So, the amplitudes associated with higher frequencies do not need to be calculated at all. (Since there are two amplitudes per frequency, one for a sine and one for a cosine, we can represent 1 s of each sound signal without audible difference by 40,000 coefficients. This is why music is digitally recorded with about 40,000 coefficients per second. In practice, a few extra coefficients are often taken into account for all kinds of reasons. For example, there are 44,100 for a CD).

For the cochlear implantation, the surgeon also has his limitations. It is undoable to attach an electrode to each of the nerve fibres. In practice, there are only 22 electrodes, and the speech processor extracts only the 22 most important frequency contributions from the incoming sound. Those 22 electrodes receive a new electric pulse every 1.25 ms (or 800 coefficients per second), and the electric current varies between 0 and 1.75 milliamps. The sound quality is therefore significantly inferior to that of a CD.

The fast Fourier transformation Also with this much more modest objective, the calculation should not be underestimated. A practical solution only came in 1965 with the fast Fourier transformation, an algorithm by the Americans Cooley and Tukey (although it is fair to say that the Prussian Carl Friedrich Gauss already used a similar algorithm in 1805 without anyone paying attention at the time). Their inspiration was to recursively split the row of incoming signals into half-length sub rows and recombine the results afterwards. This resulted in a strong reduction of the necessary computing time. (Also, Cooley and Tukey had no hearing aids in mind at first. For them, the fast Fourier transform was part of a technology to detect when the Russians were testing a nuclear weapon). As you can imagine, the fast Fourier transformation appears everywhere where signals are processed and is seen as one of the most important algorithms of all time.

The development of cochlear implants was a revolution for children who lost their hearing due to meningitis. It is estimated that 324,200 such implants had already been placed worldwide as of 2012, and the number is still increasing every year. It is a fine example of a successful collaboration between the mathematics and microelectronics of engineers on the one hand and the physiological insight of the medical world on the other hand. With the latest technologies, cochlear implants meet ever better requirements: they are small and light, and consume little energy. They don't experience any interference from technological medical equipment and mobile phones, and they don't cause any interference themselves. They can take a beating and are waterproof, easy to operate, etc. Despite the very rough reproduction of the captured sounds with only 22 electrodes, the wearers succeed very well in understanding what is being said to them. This is largely due to the fact that the recognition of words is strongly controlled by the brain. To correctly estimate pitch and timbre, important for the appreciation of music, much more detailed information about the incoming sound signal is needed. Because there are clear physical limits to the number of electrodes that can be built into the ear, current research focuses on efficient mathematical techniques to correctly pass on more detailed information with the available electrodes.

Perception is Relative, and Sometimes 100 + 100 = 106

During daily life, our body constantly performs measurements 'on sensation', whether consciously or unconsciously. We 'measure' sound intensity to adjust the volume of the radio. We 'measure' brightness to dim the light. We 'measure' how much we earn and how much storage we think we need. When the butcher hands us our meat, we 'weigh' it in our hands. We even 'measure' unconsciously and continuously how time slips by.

In 1860, the German physicist Weber worked on the border between physics and experimental psychology. He studied how people carry out all kinds of intuitive measurements and came to an astonishing conclusion: people are only able to determine *relative* differences, not absolute ones. He did this by having people pick up weights and asking whether they weigh the same or not. A weight of 100 grams required comparison with a weight of 105 grams before the subject felt the difference. For a weight of 200 grams, differences were only noticed when the second weight was 210 grams. His conclusion was that differences of up to 5% of the basic weight could not be detected by human sensation. A handy observation for, for example, shrewd bakers.

He continued to work on these observations, among others with his student Fechner, and he noticed that the phenomenon is much more general. It also applies to other senses and even to our general awareness of size. For example, doubling the sound level will always have the same effect for our ear, regardless of the basic level of sound. We make this concrete with an example. One jackhammer produces a sound wave with a sound pressure level of 2 Pascal (Pa). (The normal air pressure is 101 325 Pa. The sound pressure is the degree to which the pressure in the sound wave deviates from the air pressure in the absence of sound). Two jackhammers together give a sound pressure of 4 Pa. Four drills would give a sound pressure of 8 Pa, but for our hearing this is a doubling. The transition from one to two jackhammers gives our hearing the same sensation as a transition from two to four.

To create meaningful units in such situations, a logarithmic scale is useful. (That may sound complicated, but it simply means that we convert one quantity into another via a formula that contains the logarithm). Specifically for sound, we look at the decibel scale. We then convert the pressure in Pascal to decibel (dB) via the formula $L = 20 \log (P/P_0)$, in which L represents the sound level in decibels, P the sound pressure in Pascal and P_0 a reference pressure, here chosen to be $P_0 = 2 \cdot 10^{-5}$ Pa. Because $\log(1) = 0$, a sound pressure P_0 corresponds to 0 dB. This is a good thing because such weak sounds are not audible to us. Moreover, with this formula, we can very easily calculate that one jackhammer corresponds to 100 dB, two jackhammers to 106 dB and four jackhammers to 112 dB. Doublings lead to constant increases in the decibel scale, exactly what we wanted.

We encounter the same logarithmic scales in just about everything that has to do with perception. The division of the scale into tones is logarithmic in frequency. Richter's scale for the power of earthquakes is logarithmic. The pH-scale for acidity is logarithmic. Our eye perceives light intensity logarithmically (When a picture fails because it is overexposed, it is often because the sensors in a digital camera cannot cope with light intensity as logarithmically as our eyes can. As a result, bright light that is perfect for a human being to see is often no more than a white spot for a digital camera).

Even our sense of time is relative. It is no coincidence that time goes faster as you get older: people unconsciously compare the length of each additional year with their current age. If we could indicate years as periods of time that lasted equally long in our perception, we would also arrive at a logarithmic time scale.

Chapter 5
Google PageRank, or the Needle in the Haystack

At the beginning of 2015, according to Google, the *world wide web* consisted of 60 trillion (60,000 billion) individual pages. That number is still rising. Every day, around 3.5 billion searches are entered – around 40,000 per second. For each of those searches, a list of several tens of thousands of relevant web pages appears in less than an eighth of a second. The page you are looking for is then usually among the first ten. But how does Google do that? How does Google find the needle in the haystack?

What's the problem? To find useful pages, search engines need to know very well what content each web page contains. To do so, they use computer programs that continuously browse from page to page and store all the content on those pages in a large database. That is called indexing, and the programs that do that are called *web crawlers*. They keep track of which words appear on each page, how often they appear, and perhaps which words are close to each other. In this way they can very quickly find all the pages that might interest you. But the important work still needs to be done: in what order are the results shown? How to make sure that the most relevant pages are at the top?

The development of PageRank and the creation of Google Before Google existed, the order of the selected pages depended solely on the text on those pages. Hypothetically speaking, a search using the keywords 'Lollapalooza' could result in a top hit that shows the word 'Lollapalooza' a few thousand times in a row and that otherwise contains no other information. Not immediately the page that would make you jump for joy. So, there was an urgent need for a better system.

That system was introduced in 1996. At that time, Larry Page and Sergey Brin developed an algorithm that allows pages to be arranged according to

© The Author(s), under exclusive license to Springer Science+Business Media, LLC, part of Springer Nature 2022
G. Samaey, J. P. L. Vandewalle, *The Invisible Power of Mathematics*, Copernicus Books, https://doi.org/10.1007/978-1-0716-2776-1_5

their 'importance' on the world wide web. They did this during their doctoral research at Stanford University and called the algorithm *PageRank*, a word joke with the name of Larry Page. PageRank looks at the links that each page contains to other pages to get an indication of importance: if many pages contain a link to a particular page and those pages are important themselves, then that particular page will be considered important too. You will notice that no information is used here about the *content* of the pages. The algorithm only looks at how pages refer to each other. The content of the page is used afterwards to filter the results. Of all pages that contain the search term 'Lollapalooza' sufficiently often, the 'most important' page (the one with the highest PageRank value) will be shown at the top. Although it was clear from the beginning that PageRank delivered superior search results in all tests and demonstrations, existing search engines (such as Altavista and Yahoo) were not interested in using the technology. Page and Brin then founded their own company in 1998, which they called Google.

Where does the casual surfer pass by? The idea might be intuitively clear, but the devil is in the details: how do we determine exactly how important a page is? Page and Brin's idea is as simple as it is ingenious. They put themselves in the position of a random surfer on the internet: a person (or a computer) that is not looking for anything specific and just browses from page to page. That surfer is only allowed to visit pages by following a link on the current page. (Clearly, the rules must also allow getting out of pages that have no outgoing links, but we will come back to that later.) The criterion for impor- tance then becomes very simple: the more often the surfer lands on a specific page, the more important that page. The idea of Page and Brin is conceptually simple, but following a random surfer over the entire internet is unfortunately not really realistic. There are a lot of pages. For any single page, the probability of ending up exactly on that page is very small. This naive way of working is useful to explain the logic behind PageRank, but has no chance of actually calculating PageRank. Because the influence of accidental choices for a large internet has a very long influence on the calculated results, it would take an exasperating long time to get an accurate estimate of PageRank.

A very small Internet To see how PageRank is effectively calculated, let's take a look at a small example. We consider a very small Internet with only 4 pages, as shown in Fig. 5.1. On each of those pages we find links to at least one other page. These links are represented by arrows. Suppose we start on page 1. We can choose between two links, and so we land with a 50% probability on page 3 and with a 50% probability on page 4. If we start on page 2, we can only go to page 4, so a 100% probability. We can now look at what happens to a large group of random surfers who all start on page 1. Because the probability to go to page 3 or 4 after one step is 50%, we expect half of the group of surfers to go to page 3 and half to go to page 4. This situation is shown in Fig. 5.2. We can continue in the same way and look at what happens to the surfers on page 3 after one step. One third of them end up on page 1, another third on page 2 and another third on page 4. The 50% of surfers who were on page 4 now all surf to page 3. So 50% of the surfers are on page 3 after two steps, and 16.67% are on page 1, 2 or 4. This situation is shown in Fig. 5.3.

We can continue in this way. To avoid redrawing the figure of our network every time, we make a table in which we indicate per column what percentage of the accidental surfer is on each of the pages after n clicks. So there is a row

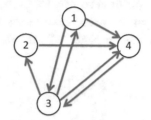

Fig. 5.1 A very small internet with four pages

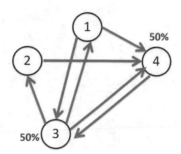

Fig. 5.2 A very small internet with four pages. In red per page is the percentage of surfers that arrive from page 1 after one click

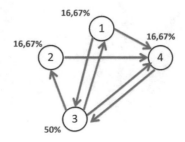

Fig. 5.3 A very small internet with four pages. In red per page is the percentage of surfers that come from page 1 after two clicks

per page, and the columns are consecutive moments in time. That table looks as follows:

	n = 1	n = 2	n = 3	n = 4	...	n = 100	n = 101
p(1,n)	0	16.67%	16.67%	8.33%		13.33%	13.33%
p(2,n)	0	16.67%	16.67%	8.33%		13.33%	13.33%
p(3,n)	50%	50%	25%	50%		40%	40%
p(4,n)	50%	16.67%	41.67%	33.33%		33.33%	33.33%

We notice that, after a while, the fraction of surfers on each page does not change any longer. The distribution of the surfers over the pages has then equilibrated. (Of course, the individual surfers keep jumping from page to page. The fact that the fraction of surfers on each page has equilibrated simply means that each departing surfer is compensated by an incoming surfer). For our small internet with 4 pages, we see that after some time, there constantly are 40% of the random surfers on page 3. So this is the most important page.

PageRank Quick Calculation... In a Week

Letting a lot of random surfers surf around Is still a lot of work. Fortunately, there is a mathematical trick to make the calculations a lot faster. To use the trick, we first need to collect all the information about our small network in a matrix *M*. Column 1 contains the probabilities (from top to bottom) to move from page 1 to pages 1, 2, 3 and 4, respectively after one click. Column 2 contains those probabilities when we start from page 2. So the matrix *M* contains in the *k*-th column row by row the probabilities to go to every page, given that you are currently on the *k*-th page. For our small network, the matrix *M* is shown on the below.

$$M = \begin{matrix} k_1 & k_2 & k_3 & k_4 \end{matrix} \begin{bmatrix} 0 & 0 & \frac{1}{3} & 0 \\ 0 & 0 & \frac{1}{3} & 0 \\ \frac{1}{2} & 0 & 0 & 1 \\ \frac{1}{2} & 1 & \frac{1}{3} & 0 \end{bmatrix}$$

In matrix *M*, the numbers are shown as fractions, but that is the same as the percentages. You can see that the sum per column is 1. This is logical, because the surfer has to go *somewhere*: the probability of ending up on one of the other pages is 1 (or 100%).

(continued)

We now try to calculate directly which fraction of random surfers will be in equilibrium on each page. We note with $p(k,n)$ the probability that a random surfer will be on page k after n clicks. Then we can calculate the chance that that surfer will end up on page k after $n + 1$ clicks. This is simply the sum of the prob-abilities that he was on one of the four pages after n clicks and then surfed to page k with the $(n + 1)$-st click. In the general case we then have.

$$p(k,n+1) = M(1,k)p(1,n) + M(2,k)p(2,n)$$
$$+M(3,k)p(3,n) + M(4,k)p(4,n).$$

The first term in that equation contains the probability $M(1,k)$ that a random surfer will end up on page k from page 1 with a click, multiplied by the fraction of surfers that were in page 1 after n clicks. So this is the fraction of surfers that will end up on page k from page *1*. The following terms are the fractions of surf-ers that end up on page k from pages 2, 3 and 4.

If we now introduce a vector $P(n)$ that contains the probabilities to land on each page after n clicks, we get the equation for the vector $P(n + 1)$ (the proba-bilities to land on each page at time $n + 1$) as a product of the matrix M with the vector $P(n)$:

$$P(n+1) = M \cdot P(n)$$

The crucial observation comes now. We are only interested in the final prob-abilities for very large values of n, when those probabilities don't change any-more. Then we obtain:

$$MP = P.$$

This equation expresses that, after infinitely many clicks, the probability of landing on a certain page does not change by clicking again. The equation above is an example of an *eigenvalues problem* and there are many methods to solve such equations efficiently. This mathematical reformulation makes the PageRank calculation much faster, but even the fastest methods used by Google still take a week to calculate! Fortunately, this is not such a big problem, because the PageRank does not need to be determined during every search.

How does PageRank stay up to date? Google calculates the PageRank of each page approximately monthly. As soon as the new PageRank is calculated, it will be distributed to all Google's servers worldwide. Because it's never fixed in advance which of those servers will handle your search, the order of the search results can fluctuate a lot during that distribution phase. Sometimes even from second to second. This phenomenon is sometimes called the

'Google Dance'. This way of working has a few important disadvantages. With PageRank in its naive form, it would take a month before new pages could be shown by Google as search results. In the early days of Google, this was also the case. Meanwhile, on top of the monthly recalculation, Google also has a daily *fresh crawl that* specifically searches for new pages. These then get a PageRank without changing the PageRank of other pages. That process is called *Everflux*. For news sites that are searched by Google News, this recalculation even happens continuously.

The secret ingredient   The only thing really missing now is the solution to a technical – but easy to understand – problem. The internet is not entirely interconnected, but consists of a large number of subsets of web pages. Each of these subsets contains webpages that refer to each other, but there are no links between the subsets so that a random surfer always gets stuck in a small part of the internet by the choice of starting page. To circumvent this nuisance, Google changes the behaviour of the random surfer: instead of always following a link, the random surfer sometimes starts over. In a fraction α of the cases, he does not follow a link from the current page, but goes to a random page on the internet. This part of PageRank is only released very vaguely by Google: the value of α is a well-kept secret and there is also no communication about the exact way in which the random new page is chosen.

Is this all there is?   Although PageRank is the heart of the Google search engine, Google also relies on other criteria for the order of the search results, in total about 200 criteria. These include some of your properties (your location, your search history, your browsing history), but also characteristics of the pages considered: age of the page, frequency of updates, correctness of language, etc. Many of these criteria are kept secret, both to stay ahead of the competition and to make sure no one can influence the search results.

The importance of fast and reliable search engines on the Internet can hardly be overestimated. Not only do they increase efficiency in our work and in the organisation of our daily lives, but they also ensure the availability of information to broad sections of the population. It is without reason that some countries try to censor Google's search results.

Despite the great success, one is still working hard to show the search results faster. After all, a small gain in speed can have a major effect on the ecological footprint of the internet. The energy consumption of our massive googling should not be underestimated. Google's total annual energy consumption is equivalent to that of 200,000 households. The energy consumption of a single search is equivalent to leaving a 60 W lamp on for 17 s. In order to display search results within one eighth of a second – the target at the start – Google spreads each search over ten data centres. In total, the intervention of about 1000 computers is required. Only the result of the fastest one is used. Google places its gigantic data centres hence in places with many cheap and natural sources of electricity. But the energy cost remains gigantic. New algorithms, based on improved mathematical insights, can therefore have a major impact on energy consumption. Even a very small gain in speed can make a big difference because of the large volumes.

Chapter 6
Fun and Game Theory

To take off the pressure or for teambuilding, few things are more relaxing than a fun and exciting board game. But which game elements determine whether a game is exciting? And how can a game remain exciting, even if you have already played it quite a few times before?

What's the problem? In a good board game, it is important that the outcome not only depends on luck but can also partly be determined by the choices you make as a player. A game also turns out to be more fun if beginners have a chance to win but advanced players can benefit from experience at the same time. Moreover, ideally every game instance should play out differently and there should be different good strategies to choose from. Under these circumstances, the strategy that leads to victory will also depend on the choices of the opponents. But how can game designers take care of all that? How do they ensure the right balance between luck and skill?

The Game of Goose is elaborate gambling When talking about coincidence and strategy, the games that immediately come to mind are quite extreme in this respect. The well-known Game of Goose is at one extreme. The course of the Game of Goose is entirely determined by a sequence of dice throws and is therefore just a long drawn out game of luck. To find out how the game pans out and to fine-tune the rules of the game in such a way that even during the game it is not quickly clear who is going to win, a mathematical analysis can be helpful. To do this, we present the board as a network. Each box is then a node. The possible transitions at dice throws are indicated by arrows. We play with one die for the simplicity of this explanation. From square 1, for example, you can move to 2, 3, 4, 5, 7 and 12 (because there is a rule on square 6 that allows you to move to 12). In this way we can calculate all possible paths

© The Author(s), under exclusive license to Springer Science+Business Media, LLC, part of Springer Nature 2022
G. Samaey, J. P. L. Vandewalle, *The Invisible Power of Mathematics*, Copernicus Books, https://doi.org/10.1007/978-1-0716-2776-1_6

from start to finish. This leads to a number of surprising observations. For example, in a game between two players, the starting player has a slightly better chance of winning. However, the game has been designed in such a way that it is intuitively difficult to estimate how much you are ahead or behind your competitors. An example of such a rule is the well. If you end up in the well, you'll have to wait for another player to come along to take you out. (The prison does about the same thing.) This should guarantee the excitement. Unfortunately, it doesn't seem to be that bad to end up in the box with the well: your chances of winning are hardly affected by it. And with two players there is about a 23% chance of a tie (one player being in the well and the other in prison). Even with all these problems, the main problem of the Game of Goose is much more basic: players can't make any choices and just observe the outcome, so the game doesn't remain interesting for long.

Antoine Gombaud's gambling game and the creation of the probability theory

Antoine Gombaud was a French nobleman and writer at the beginning of the seventeenth century, and it seems that he often did not know what to do with his time (and money). He was in fact a firm gambler and invented his own games. In a first version of his game, he proposed to roll a die four times. If at least one of the four outcomes was a 6, he won. Otherwise he lost. Gombaud had figured out (but did not tell the others) that he had a 4/6 (66.7%) chance of winning with four rolls. He also won regularly, so often that his friends didn't want to play with him anymore.

That's why he proposed a second version of his game: he would now throw two dice 24 times and win if there was a 12 in between. According to his calculations, this gave him the same winning odds (namely 24/36) and he hoped that his friends would not be as smart. Unfortunately, he started to lose hard. He soon realised that there had to be a mistake somewhere in his plan.

What does a seventeenth-century nobleman do in such a case? Right, he wrote a letter to the most famous mathematician of his time – in this case Blaise Pascal – and asked him to check his calculations. Unfortunately, Blaise Pascal was also a bit overwhelmed: he didn't see immediately where Gombaud went wrong and in turn contacted Pierre de Fermat. The Dutchman Christiaan Huygens was also called in at some point. The correspondence that these three mathematicians conducted after this led to the start of probability theory.

In a first phase – before drawing up a general theory – Pascal and Fermat tried to discover Gombaud's fallacy. The first conclusion is clear: with Gombaud's reasoning, he would have had a 100% chance of winning the first version of his game if he would be allowed six dice rolls. That's clearly not true! You can certainly remember situations where you needed a six in a game that simply didn't

(continued)

come, no matter how many times you threw. The mistake is even worse: with that reasoning, if you were allowed to roll more than six times, you'd have *more than a 100%* chance of winning! There clearly is something fundamentally wrong with Gombaud's reasoning. Actually, you expect your probability to throw a six to increase with the number of times you are allowed to throw and to approach 100% when you are allowed to throw an infinite number of times.

Interestingly enough, the first step Pascal and Fermat took was to turn the problem around. They calculated the chance of *not* winning: the probability in the first game to not throw six in every turn is 5/6. After that, they noted that the composite odds are the product of the individual odds, not the sum. The chance of loosing is then $(5/6)^4$ or 48.2%. It follows that Gombaud had a 51.8% chance of winning with his first game. They first verified this result by writing down all possible outcomes and counting the winning odds; the proof only followed later. For the second game, the probability of not winning with the same reasoning turned out to be $(35/36)^{24} = 50.8\%$. That's why Gombaud lost this game more often than he won!

Nothing is coincidence in chess At the other extreme we have chess. In chess, every move is a conscious choice of one of the players. Randomness or luck is therefore out of the question. Moreover, both players have exactly the same weapons – except for the right to start. It is then basically possible to calculate all possible sequences of moves from any position and see who ends up winning. In theory, a computer can then calculate its 'best move' as the one that will yield a win in most of the possible outcomes.

In practice, the number of possible positions is unfortunately far too large to actually do this: there are about 10^{40} possible ways to arrange all chess pieces in a regulated manner. The number of possible chess games that could be played is estimated at about 10^{120}. In comparison: the number of atoms in the observable universe is about 4×10^{80}. Deep Blue II, the computer that could defeat world champion Kasparov in 1997, calculated 200 million proposal moves per second and could still only think 12 moves ahead. A good chess player can think forward much faster because really bad moves don't even occur to him. That saves a lot of calculations. That's why chess is a game in which a novice has no chance of beating an experienced player with a bit of luck.

And what about four-on-a-row and checkers? Another game in which coincidence plays no role is four-on-a-row. However, the number of possible games four-on-one-row is much smaller than the number of possible games of chess: there are exactly 4,531,985,219,092 ways to fill a four-on-one row 6×7 board with stones. In 1988 James Dow Allen and Victor Allis (separately) calculated all possible games. This shows that the starting player can always win if he starts in the middle. Although there are no simple rules for winning moves, the entire database with all possible four-on-a-row games exists as an app for smartphones. Four-on-a-row is therefore called a solved game. There are also perfect strategies from the starting line-up for checkers. However, these are not winning strategies: they only guarantee a draw. At the World Championships, this problem is avoided by randomly selecting an opening for each game. The game then starts from a position in which three moves have already been made that a sensible player might never try. At checkers, there are 156 of such openings possible, and they even have been ranked according to difficulty (for both black and white).

Attacks in Risk: a combination of strategy and luck Most games today combine aspects of strategy and luck. In Risk, for example, countries are attacked by armies. The strategic aspect is to choose who you attack and when. The outcome of the attacks themselves is determined by chance. Both the attacker and the defender roll a number of dice (maximum three) and the pips are compared two by two. The attacker has an advantage because he can always throw more dice than the defender. This advantage is compensated by the fact that the defender wins in case of a tie. Because battles between large armies often take place in different phases, it is not so immediately clear which of the two parties ultimately has the biggest advantage when both armies are about the same size. A thorough mathematical analysis was needed to show that the attacker has a slight advantage in that situations, but only if his army consists of five cubes or more. If the armies are large enough (from about 11 cubes), the attacker has the best chance of winning, even with a slightly lower number of cubes than the defender.

Information asymmetry, game balance and game theory Many games provide additional strategic elements. This can be done, for example, by ensuring that not everyone has the same information. In many card games, this is regulated by giving each of the players a limited number of cards that are unknown to the others. Often players also have different roles, which means that the same card has a different use for each of the players. The well-known card game Citadels is a good example of this. It is then very important to make sure that the game is balanced: no individual card should be so strong that it completely determines the outcome of the game. Because it is very difficult to get such a balance, drawing up game rules is accompanied by quite a bit of mathematical analysis. A crucial aspect of that analysis is estimating optimal strategies. When a game is designed in such a way that players can formulate an optimal strategy for themselves without being dependent on the behaviour of the other players, the game quickly loses its appeal. As soon as the players realise this, everyone chooses that strategy and the outcome of the game is fixed from the start. A game only becomes interesting when you have to adapt your strategy to your opponent's strategy, and especially when information about your opponents becomes available only gradually. There is a whole branch of mathematics dedicated to this kind of analysis that is called game theory.

Working together or opposing: the deer hunt

Deer hunting is an analogy invented by the French philosopher Jean-Jacques Rousseau (1712–1778) to reflect on collaborative behaviour among people. It also often occurs in one form or another in games. Deer hunting is about the choice between working together for a greater purpose or going your own way for quick profit. The situation is as follows: two friends (Alice and Bob) go hunting together and have the choice between two prey: a deer and rabbits. The deer is more difficult to catch: that only works if Alice and Bob work together. Alice and Bob can easily catch rabbits on their own, but if they both do so, they have to share the loot.

Both Alice and Bob now have to make a choice: hunt the deer or rabbits. Because they both choose independently, there are four possible outcomes. We put them in a table. In the columns we put the two choices of Alice, in the rows the two choices of Bob. In the corresponding four boxes we place the revenues for both Alice and Bob:

	Alice hunts deer	Alice hunts rabbits
Bob hunts deer	Alice: 3 | Bob: 3	Alice: 2 | Bob: 0
Bob hunts rabbits	Alice: 0 | Bob: 2	Alice: 1 | Bob: 1

We see that the yield is highest when Alice and Bob hunt the deer together. Then they both get a reward of 3. That is clearly the best strategy. It is risky because it requires both Alice and Bob to trust that the other will cooperate. However, if one of them chooses to hunt a rabbit, he gets all the rabbits (a reward of 2); the other one gets nothing. When they both hunt rabbits, they have to share the rabbits. They each get a reward of 1. The latter is a risk-free strategy, in which both Alice and Bob are satisfied with the certainty of a small reward.

We can now find out which situations are in 'equilibrium'. We call a situation in equilibrium when both Alice and Bob have no reason to change strategy unilaterally. That definition was invented by John Nash in 1950, and is therefore called a Nash equilibrium. The risky collaborative strategy is clearly a Nash equilibrium: both Alice and Bob lose out when they decide to unilaterally start hunting rabbits if they were hunting deer together before. The situation where Alice hunts deer and Bob hunts rabbits is not a Nash equilibrium: it is advantageous for Alice to adapt her strategy and to also start hunting rabbits.

Although this may go against intuition, there is also a Nash equilibrium when Alice and Bob both hunt rabbits: a unilateral decision to start hunting the deer can only lead to loss (and more rabbits for the other).

This simple game explains why, when there is a lack of confidence, people can choose strategies that are clearly not optimal, both in games and in serious cir-

(continued)

cumstances such as business deals or political negotiations. John Nash himself and his wife were killed in a car accident in early 2015. Ironically, the circumstances of his death can be described by the above game. Although it would have been optimal for the couple to put both their seatbelts on, they weren't wearing them. You can argue that the couple was stuck in the least favourable Nash equilibrium. When either of them would rather die than continue living alone, neither tend to put their seatbelts on unilaterally. However, it is rather inappropriate to speak of a 'risk-free' strategy in this case.

Game theory is not only (and not mainly) about drawing up good rules for games. Because we often encounter situations in everyday life where we have to make choices, game theory also has applications in domains such as politics, economics, biology and computer science. In economics and politics, game theory is used to assess the strength of negotiating positions and to make decisions. In sociology it is used to formulate theories to understand how people make choices. And game theory even comes in handy when complying with traffic rules or for describing the mating behaviour of certain animal species. Meanwhile, game theorists have already collected 11 Nobel Prizes, mostly in economics. The American John Nash, who became schizophrenic around the age of 30 and whose life was filmed in 'A beautiful mind', is probably the best known.

Part II

Mathematics in the Workplace

Chapter 7
High Winds on High Mills

To combat climate change, massive efforts are being made to generate renewable energy. For example, the EU has a target of achieving 32% of its energy needs from renewable sources by 2030, with a potential upwards revision in 2023. Wind would then be one of the most obvious sources. Unfortunately, there isn't always a lot of wind, and it doesn't always come from the same direction. Can we make wind a reliable source of energy? And is our power grid ready for such a boost of green energy?

What's the problem? Since the industrial revolution, the earth's prosperity has increased spectacularly, mainly because energy became available quickly and cheaply. However, burning fossil fuels has negative consequences for the climate. Nuclear energy in turn produces radioactive waste and has already given rise to disasters such as Chernobyl and Fukushima. That is why we are increasingly focusing on renewable energy sources to meet our (still growing) energy needs. Unfortunately, most renewable sources – such as sun and wind – are not constantly available, and the amount of energy generated can vary greatly over time. This gives rise to a lot of interesting questions: How can we generate wind energy most efficiently? And how does the electricity grid deal with variable production and consumption? (Fig. 7.1)

Design of efficient wind turbines The power of a wind turbine is mainly determined by the blades. The larger they are and the faster they rotate, the more energy they generate. This is why there is a strong tendency to build ever larger and higher wind turbines, currently sometimes with a mast over 100 m high and blades with a length of 60 m.

G. Samaey, J. P. L. Vandewalle, *The Invisible Power of Mathematics*, Copernicus Books, https://doi.org/10.1007/978-1-0716-2776-1_7

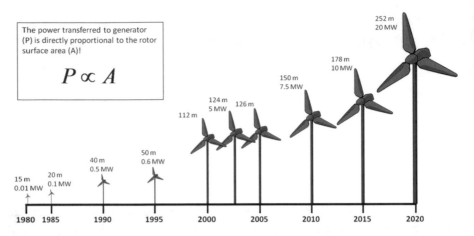

The power transferred to generator (P) is directly proportional to the rotor surface area (A)!

$$P \propto A$$

Fig. 7.1 Evolution of the size and capacity of wind turbines since 1980. The power supplied is proportional to the surface area of the blades

In addition to the size of the turbine, the shape of the blades is also very important. The shape determines how the air flows around the blade and, consequently, how fast it will rotate. That is why this shape is optimised in such a way that as much energy as possible is transferred from the wind to the blades. However, there are limits. As early as 1919, the German physicist Albert Betz proved that a maximum of 59.3% of the energy present in the wind can be captured by a turbine. He did his calculations for classic windmills, such as those often found in Flanders and the Netherlands – at that time there was still a long way to go to get electricity from wind energy. The reason for the limit is that the air behind the blades cannot come to a complete standstill because it still has to get away to make way for the oncoming wind. Current wind turbines achieve about 75% of that maximum efficiency. The extra loss is mainly caused by friction in the rotating parts of the turbine that convert the mechanical energy into electrical energy.

The optimal shape of the blades also depends (unfortunately) on the wind speed. And wind speed can, as already said, vary quite a lot. We therefore also have to calculate what changes in the generated energy we can expect with changes in wind speed. In general, it is better to take a blade that works reasonably efficiently in a wide range of weather conditions, instead of a blade that is most efficient for one specific set of weather conditions. Over a longer period of time, such a choice produces a better yield than a blade that generates just a little bit of extra electricity for a specific wind speed.

Safe and quiet In addition to efficiency, safety is also an important design criterion. Nobody wants the blades of the wind turbine to break off and crush someone on the ground. Depending on the material they are made of, the blades deform under the force of the wind. To avoid breaking the blades, the mechanical forces exerted by the wind on the blade must be well spread out. If the forces are concentrated at a certain location, there is a high probability that the blade will break at that location. Even if the wind speed is not always high, the tops of the blades rotate quite fast, especially for the largest turbines. This results in a very complex flow pattern and consequently a very complex distribution of mechanical forces inside a blade.

Noise nuisance is also an important point of attention in the design of the shape of the blades. Especially in densely populated areas it is important that the rotating blades do not make too much noise. In addition, it turns out that the noise is ideally 'white noise', a kind of background noise in which both high and low frequencies occur, a bit like a radio that produces a garbled sound and is not tuned to a specific station. It turns out that people can ignore this kind of noise much more easily than a single precise whistling tone.

Kites instead of windmills?

In addition to the size of the blades, there is a second important reason for wanting to build higher wind turbines. In higher layers of air, the wind also blows much stronger and more constantly. A major problem with the increasing size of wind turbines is that they require so much material that they are no longer profitable in the long run due to the high production costs. A 100-metre high wind turbine requires 200 tonnes of steel, and 18 tonnes of fibreglass is used for 60-metre long blades. In addition, power increases with the *surface* of the blades, while it is the *volume* of the mast and the blades that determines how much material is needed. Unfortunately, as turbines become taller, the volume of the mast increases much faster than the surface of the blades. At the same time, about 60% of the energy is taken from the thin outer 30% of the blades. The rest of the construction – that mast and the rest of the blades, which contain the majority of the mass of the turbine – only serves to keep that part in place.

This observation leads to a radically different way of exploiting wind energy. We could actually try to make a kite instead of a windmill. We then only need to keep the outer 30% of the blades and connect them to the ground through a long cable. When the kite flies away, we generate electricity. We use some of that electricity to pull it back.

The great difficulty is to keep the kite in the air without human intervention. If you have flown a kite as a child, you know that you have to keep a close eye on the movement of the kite and make appropriate adjustments by pulling on the wires. The latter is done with the kites by computer programs that calculate where the kite is currently going and then adjust it when necessary to keep it in the air.

Virtual wind turbines in virtual weather conditions Efficiency, safety and noise are therefore the three main focus areas in the design of wind turbines. The fact that computer simulation is useful in this respect is not so difficult to understand: the only alternative to simulation is an experiment in a wind tunnel. However, today's wind turbines are far too large to fit in a wind tunnel. So, we can only physically test reduced scale models. Unfortunately, we cannot simply convert the behaviour of a scale model to a realistic size. (We can do that for their energy efficiency, but not for noise production and safety.) In mathematical terms, this is due – briefly – to the non-linear interactions between the wind, the distortion of the blades and the noise production. Computer simulations are therefore not only useful, but also crucial. There is no alternative.

From wind turbines to wind farms Wind turbines are usually put in groups, called wind farms. An additional advantage of computer simulation is that we can also calculate the behaviour of entire wind farms. There is certainly no room for large-scale wind farms in a wind tunnel. Wind farms have additional complications, such as the optimum positioning of the wind turbines with respect to each other, which strongly depends on the wind patterns at the location where the wind farm is to be installed. Each wind turbine has a strong influence on the local wind flow. It must therefore be ensured that the wind turbines do not catch each other's wind because that would render part of the investment useless. The crucial question then is: how should we install the wind turbines so that the energy they generate together is at its maximum? This leads to a very big optimization problem in which the generated energy is written as a function of the positions and orientations of the wind turbines. We then look for those positions and orientations that maximise the amount of energy generated.

Some scientists even go a step further and investigate how these wind farms influence the weather. For example, researchers at Stanford University – in a rather speculative article – even calculated that Hurricane Katrina could have been mitigated by a wind farm before it landed in New Orleans. However, to avoid the damage caused by that hurricane, 78,000 wind turbines would have been needed, which might be a bit too much to be practical.

Turbulence in the air When simulating wind turbines, a mathematical model for the wind flow around the rotating blades must be combined with a model for the deformation of the blades themselves. The simulation must take into account variable and sometimes unexpected wind speeds and wind directions. Preferably in a way that statistically corresponds to the weather at the place where we want to install the wind turbine. Everything then has to be linked to an acoustic model if we also want to map the noise generated, and to a local weather model if we are dealing with a wind farm.

One of the major computational problems in such simulations – besides the simple fact that so many different types of physical phenomena are involved simultaneously – is the occurrence of turbulence in the airflow. Typical of turbulence is that the flow always contains a large number of vortices (or vortexes). Each of these vortices again contains much smaller vortices, and so on. Because computer simulations rely on rectangles of finite size and smaller phenomena cannot be made visible in the calculations, we have to create a mathematical model for the effect of the invisible (because too small) vortices on the airflow between the rectangles. Simply neglecting the small vortices – assuming they are not there–is in fact a good recipe for very inaccurate results.

New energy on old grids Since the end of the nineteenth century, our electricity grid has been gradually expanded based on the situation at the time: electricity is produced in a limited number of large power stations, at a fixed rate, and distributed to a large number of users. With the advent of renewable energy, the grid is now exposed to requirements for which it was not designed: the amount of electricity generated by wind power can vary greatly over time. Moreover, with the massive emergence of solar panels, we are potentially all (small) producers. To enable two-way flow of electricity, the existing electricity grid needs to be adapted. After all, the thickness of the wires must be adapted to the power they have to transport. Otherwise the connection will fail. To achieve this, sensors are added to the network at a rapid pace to gather information about local power consumption and local production. This information is then used to optimise maintenance and adjustments to the network, and above all to ensure that there is no overvoltage or undervoltage anywhere. The network becomes unstable if a higher voltage or current than the cabling can handle is generated somewhere. At that moment it must be possible to decide within seconds to switch connections on or off so that no cascade occurs that leads to a massive power failure. To exclude such risks as much as possible, agreements are made that at first glance seem strange. Steel companies, for example, use a lot of energy. (We look into this again when it comes to accurate forecasting in Chap. 13). The price they pay for their consumption often depends not only on the amount of energy they use, but also to a large extent on the accuracy with which they can predict how much they will need. In many countries, such as Belgium, owners of solar panels also need to register their panels, at the risk of high fines if they don't do this, because local electricity producers could upset the balance of the power grid.

Make money using electricity? Ultimately, the stability of the electricity grid depends on the balance between supply and demand. As every economist knows, price is an excellent way of achieving that balance. It is therefore not surprising that a great deal of research is being done into ways of steering the power grid on the basis of continuously evolving prices. This can be interesting when you don't need electricity all the time, and certainly when you are able to choose your moments. Charging electric cars is a good example of this. In a business context, you can think of the cold stores in which supermarket chains store their fresh products. Companies such as the Flemish REStore ensure that such large consumers of electricity buy their electricity at times when there is an oversupply on the network, and the price is therefore very low. Sometimes the price can even be negative: in that case you are even paid for buying electricity.

The transition to an economy based on renewable energy is one of the biggest challenges of the twenty-first century. It is essential in the fight against climate change, and renewable energy is also crucial for geopolitical reasons: it allows us to be more autonomous in our energy supply, reducing our dependence on other countries. However, to achieve this transition, many mathematical problems still need to be addressed, both in the production and consumption of energy.

Chapter 8
Reducing Factory Noise

Working in an excessively noisy environment causes stress and fatigue, not to mention hearing damage. Noise can also cause unsafe situations because employees do not hear each other shout. For these reasons, the maximum noise level at the workplace in Flanders is limited by law to 87 dB. In factories, most noise is caused by loud machines. Can we take the noise level of the machines into account in the design phase? And how much extra profit can we expect from such optimization?

What's the problem? Machines make noise because they vibrate. The faster they rotate, the stronger they vibrate and therefore the noisier they are. In the textile industry, for example, automatic looms are responsible for a lot of noise due to the large number of moving parts. For a factory, the easiest solution is often to slow down the operation of loud machines, but of course this goes at the expense of productivity. So there is a clear interest in designing quieter machines. While textile production is gradually disappearing from Western Europe, the Flemish company Picanol is still a world player in the production of new looms, a high-tech sector that requires a great deal of mathematical optimization. Because the word 'loom' has a somewhat old-fashioned connotation, manufacturers prefer to speak of 'weaving machines' to highlight the high-tech character of their product. Picanol makes its weaving machines quieter by strategically attaching weights to the loom in such a way as to counteract vibrations. But what weights should we use? And where should they be hung?

A routine job: balancing car wheels Before we talk about the weaving machines in more detail, we first look at a simpler problem to understand the physical and mathematical principles: balancing car wheels. Balancing is

G. Samaey, J. P. L. Vandewalle, *The Invisible Power of Mathematics*, Copernicus Books, https://doi.org/10.1007/978-1-0716-2776-1_8

important because the wheels do not have exactly the same weight everywhere along their circumference. As a result, the wheel is not perfectly circularly symmetrical around its centre, and the resulting imbalance creates violent vibrations as soon as we start driving a little faster. These vibrations are eliminated by measuring weight differences along the circumference and compensating for them by placing appropriate counterweights. With a well-balanced wheel, there is less vibration (and therefore less noise) while driving and road contact is better. This is better for safety, driving comfort and fuel consumption. It also causes the tyres to wear out more slowly.

Vibrations in weaving machines A similar problem arises with professional weaving machines. There the vibrations are not caused by rotating parts, but by grippers that move back and forth extremely quickly with the thread being woven. The grippers are pulled by a frame which itself also vibrates. It is extremely difficult to map all these simultaneous vibrations – much more difficult than measuring imbalance on car wheels. This is because so many different parts move back and forth at different masses and speeds. Also, there are no devices to measure the imbalance and to determine the location and size of the counterweights, as was still possible for wheel balancing. So we will have to invent something else. In doing so, we arrive fairly quickly at mathematical optimization.

The optimal placement of a single weight We start with the placement of a single weight, with a predetermined mass. As soon as we know the relation between the position x of the weight and the amplitude of the vibrations of the loom, $f(x)$, we can make a figure on which we plot the amplitude $f(x)$ as a function of that position x. We can then see for which position the vibrations have the least effect. In the left figure it is for position b, in the right figure it is for position d. To be able to draw that figure, we must calculate the amplitude of the vibrations $f(x)$ for many values of the position x. For a large loom this can require a lot of calculation, even if we leave that work to a computer. It would be much better if we looked directly for the minimum. For example, we should only calculate $f(x)$ for values of x for which we have a reasonable chance of improvement over the current best position.

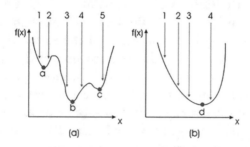

(a)   (b)

Descending into a mountainous landscape Let's see what happens if we can also vary the mass of the weight. So now we have two parameters: the position x of the weight and its mass y. We again determine the function $f(x,y)$ that represents the amplitude of the vibrations as a function of the position x and the mass y of the weight. To continue, we use a metaphor. We interpret x and y as position coordinates (longitude and latitude) and pretend that $f(x,y)$ is the height of a mountainous landscape. Now we look for the lowest point in that landscape. Of course we can walk around the entire landscape and note the altitude everywhere, but that takes quite a long time. That is the equivalent of what we did before when we drew the graph of the function $f(x)$ to find its minimum. In two dimensions, we then obtain an elevation map, a map that shows all points with the same height as elevation lines.

If you don't need the entire elevation map, but just need to find the lowest point, it is much quicker to just walk down according to the direction in which you notice that you are going down the fastest. By scanning around your current position with your foot, you can even do that blindfolded (if you are careful). When we then get to a point from where we can only leave by going up, we consider that to be a minimum. Mathematically 'we walk blindfolded down' by looking locally for the direction with the strongest descent – that is the direction in which the derivative of the function $f(x,y)$ is most negative – and then taking a step in that direction. Methods of this kind have existed for a very long time to find minima of functions. Isaac Newton (1642–1726) had already developed such a method, but only for polynomials, long before the theory of functions and derivatives in its present form was ready.

And now for real weaving machines... Before we can use this type of optimization method in the design of new weaving machines, we still have a number of practical difficulties to overcome. To begin with, we cannot accept solutions where the mass of the weight is negative. Moreover, we do not place a single weight, but several tens. The algorithms then don't 'walk around' in a normal mountain landscape, but in a mountain landscape in a space with several tens of dimensions. In addition, we can't just hang the weight in the air, but we have to hang it on the frame of the loom. There are still many practical arguments we have to take into account, such as limitations on the total mass of the counterweights, cost and available space. Finding a good compromise between all these requirements is not easy and requires not only a solid mathematical knowledge but also specific expertise on the origin of vibrations. Fortunately, this type of work is carried out in teams, with the various team members each thoroughly mastering one aspect of the problem.

Make problems convex to solve them easily

An annoying complication while optimising is the observation that sometimes there are multiple (local) minima in the function we are looking for. You can already see that in the one-dimensional case on the figure above. On the figure on the left, you can also only move from positions a and c by going up, and that while point b is even lower. In the two-dimensional case, we often have many local valleys as well. You can imagine that visually by realising that rain that lands on the ground also flows away in the direction of the strongest falling. There are then several local minima if there are several puddles. On the right figure, we don't have that problem: the minimum d is the only minimum. If we find that – and we can guarantee that we do so with the methods described above – we are guaranteed to be ready. Functions that have only one minimum (like the one on the right figure) are called *convex*.

If the function that we try to minimise is not convex (and therefore has several local minima), all methods to find those minima will get stuck in a local minimum, without knowing if we have reached the global minimum. A classic way to solve this is to start searching the minimum from many different starting positions. That of course requires a lot of calculation. It is much better to formulate the problem in such a way that it becomes convex. So we do not invent a new method to calculate the solution, but adapt the problem so that it becomes convex. Of course, we must not change the solution. That sounds like pure magic, but it turns out to be possible surprisingly often.

Faster weaving machines can compete better on the competitive international market. In addition to the noise pollution caused by excessive vibrations, there is another major limitation on the speed of these machines: the vibrations create strong forces on the highly dynamically loaded parts of the loom (such as the grippers and frames). These forces lead to rapid wear and tear and can even cause breakages. At the Belgian company Picanol, convex optimization was used in the development of the commercial machine OptiMax. The results were spectacular. The total unbalance force on the machine foundation was halved at 750 impacts per minute. The OptiMax machines are relatively quiet and with weaving speeds of up to 750 wefts per minute they were also the fastest of their kind when they entered the market in 2008. A world record for the Belgian industry, which can therefore be more competitive on the international market.

The Perfect Engine Sound

Noise is not only a problem with machines in factories. A lot of effort is also made to reduce noise pollution in aircraft, for example, in the design of the shape of the exhaust pipe of the jet engine. This is important for the quality of sleep of people living near airports, and for the efficiency of transport companies because quieter aircraft are allowed to take off later in the evening.

Cars, on the other hand, have a completely different dynamic, one that at first sight may raise our eyebrows. As internal combustion engines become more and more efficient, they also become much quieter. This gives some potential buyers the (false) feeling that the engine is less powerful. Many brands (such as Ferrari) solved that by designing special exhaust pipes that make exactly the sound you expect from a car of that brand. That too is a rather complex mathematical optimization problem.

Meanwhile, even that is no longer enough. To improve driving comfort, cars are also being increasingly insulated against outside noise. This makes conversations easier, and the radio in the car can be heard better at a lower volume. Unfortunately, the engine noise is also kept out, even if it is only made for the pleasure of the driver. Most brands nowadays solve this problem by recording the sound made by the engine and playing back the speakers. In Volkswagen this system is called Soundaktor, in BMW it is called Active Sound. The step to synthetically generated sound is then not that far-fetched. Who knows, maybe within a few years your Prius will sound like a thoroughbred Ferrari.

The latter is not even as nonsensical as it sounds at first. There are studies that show that cars that are too quiet can be dangerous for crossing pedestrians. An artificial engine noise could be a step forward for road safety in electric cars, especially for slow-moving cars (below 30 km/h) that hardly make a sound at all. At higher speeds, the sound of the tyres and the wind around the car becomes dominant, so the problem of the car being too silent disappears. Extensive tests have already been carried out with different types of noise. The additional safety has to be weighed against the annoyance of the additional noise. Artificial beeps, such as those from a parking sensor, turn out not to work so well and are, on top of that, very annoying. In the end, an artificial engine sound appears to work best. This has all the characteristics we expect as a pedestrian: it plays continuously and becomes louder when accelerating. Moreover, we have gotten used to it, so we quickly realise what the exact danger is.

There are many more fun anecdotes to tell about car design: the door of a car, for example, is designed in such a way that when it slams, the sound creates a feeling of solidity. That sound is the first a potential buyer hears when he wants to test drive – even before the engine starts! The importance of a solid door sound became apparent when car makers started incorporating reinforcement bars into doors. To avoid increasing the weight of the car, the rest of the door was made lighter. As a result, the door made less noise when closing, something that was not interpreted correctly by potential buyers. They thought that the car had become less sturdy, while it was just the other way around. The simplest solution turned out to be to design a door that would intentionally recreate the solid, but superfluous, sound.

Chapter 9
Virtual Architecture and Engineering

New airplanes and cars are extensively tested before they are allowed on the market. Buildings such as bridges and football stadiums also have to comply with a whole battery of safety regulations. It is therefore important to build and test a number of prototypes during the design phase. But can we speed up the design process by performing some of these tests virtually? And can these virtual tests be trusted?

What's the problem? Bridges can collapse. Planes and cars can crash. Riots can break out at a football match, and even if they don't, there can be sudden overcrowding when people are trying to rush their way out of the stadium. In order to guarantee the safety of a structure or vehicle in all circumstances, its creator must consider possible disaster scenarios or accidents as early as the design stage. And preferably before the product is put on the market. If problems are discovered too late, this causes huge costs, not to mention negative publicity. Just think of the failed moose test in 1997 in which the brand-new Mercedes A ended up on its side in a sudden evasive manoeuvre. To avoid that kind of debacles, prototypes are usually built and tested before a new product comes on the market. But do we always have to build and test prototypes during the design phase? Or can the design be car   ried   out   much   faster   and cheaper with computers?

The wind makes bridges tremble Bridges not only have to withstand the load of traffic passing over them, they also have to withstand the wind. This is best looked at before the bridge is actually built, because once it is there, little can be done about it. But what exactly is a prototype for a bridge? Once the bridge is built, it is built, isn't it? Fortunately, we can learn a lot about a bridge from a scale model, large enough to be faithful, but small enough to be easily

G. Samaey, J. P. L. Vandewalle, *The Invisible Power of Mathematics*, Copernicus Books, https://doi.org/10.1007/978-1-0716-2776-1_9

Fig. 9.1 The Tacoma Narrows Bridge began to swing so violently in the wind in November 1940 that it collapsed

tested. To study the effect of the wind, the prototype should fit in a wind tunnel, a test setup where air is blown around an object to investigate its aerodynamic properties. That these tests are important is illustrated by the incident with the Tacoma Narrows Bridge in 1940. That bridge in the American state of Washington was designed to withstand storms with wind speeds of 160 km/h. Even during construction the bridge wobbled slightly, apparently without major problems, making it a local attraction. However, on 7 November 1940 – only a few months after its opening – it suddenly began to swing violently under the influence of a strong breeze with wind speeds of barely 65 km/h (see Fig. 9.1). After a while, the oscillations became so violent that the bridge broke and collapsed.

Although there was still traffic across the bridge until just before the collapse, fortunately there was only one victim to mourn: the dog Hubby, who had become too scared to leave the car and even bit his owner that tried to save him. When they designed the Tacoma Narrows Bridge, the engineers had no idea that such a gentle breeze could play such a major role. Testing with scale models in wind tunnels wasn't common at the time either. It is precisely thanks to research into the causes of the collapse of the Tacoma Narrows Bridge that the importance of testing with wind tunnels has become clear – so clear that they subsequently became a legal requirement.

The catastrophic consequences of resonance In the 75 years since the collapse of the Tacoma Narrows Bridge, much research has been conducted into possible causes for this fiasco. At first it was thought that resonance was the cause. Resonance – known from a tuning fork – is a phenomenon in which an object vibrates very strongly when a force acting on it excites the natural frequency of the object. For a good tuning fork, this can be done with a hard tap. The resulting vibration then results in a pure music tone 'A' or 'la' in the solfege. A bridge also has such a natural frequency, and a hypothesis was that the wind at the Tacoma Narrows Bridge caused vibrations at that natural frequency. This reasoning did not come out of the blue. Collapses due to resonance had already been noted a number of times. The most spectacular collapse occurred on 12 April 1831 in England, with the Broughton Suspension Bridge. The bridge appeared to have a natural frequency that just matched the rhythm of the regiment of soldiers marching there that day. Their footsteps caused violent vibrations which led to the collapse of the bridge, fortunately without loss of life. The incident led to a new military directive: from then on, soldiers were no longer allowed to march on bridges, but had to walk freely. Unfortunately, this measure turned out not to be enough to prevent all possible problems. More recently, the Millennium Bridge in London had a similar problem when the crowd spontaneously started to synchronize due to the large crowds. Probably resonance arose because the people stepped at a frequency close to the natural frequency of the bridge, and they unconsciously adjusted the rhythm of their footsteps slightly in order to step in perfect synchrony with the vibrations of the bridge.

Not everything that goes wrong is due to resonance Now that we know we have to pay attention to resonance, the effect is fairly easy to prevent. It suffices to ensure that the bridge has a natural frequency that is not generated by the forces acting on it. Unfortunately, with the Tacoma Narrows Bridge resonance was not the biggest problem. There was another problem, in particular aeroelastic flutter. In a nutshell (and with some oversimplifications), the following happens: first there is some wind under the bridge that pushes the bridge up a bit. Gravity pulls the bridge back down, but too hard. This causes the bridge to end up a bit lower than where it started. Now the wind goes over the bridge and pushes the bridge further down, until the suspension pulls it up again. When the suspension pulls the bridge up a bit too hard, the bridge ends up higher than its original position. Now, more wind goes under the bridge than initially. The cycle repeats itself, but more violently than before. This whole repetition creates a self-reinforcing interaction between the vibration of the bridge and the wind, in which the bridge oscillates more and more until it breaks. The exact nature of this interaction is highly dependent on the size of the bridge, and is therefore difficult to predict from scale models. Instead, the bridge can be modelled on a computer using a technique we call the finite element method. This method divides the bridge into a large number of small pieces, each of which can move under the influence of the wind. These pieces are of course connected to each other and exert forces on each other to stay together. A precise mathematical description of these forces is based on differential calculus. Based on that finite element model, the natural frequencies of the bridge can be calculated.

Aircraft wings also vibrate The same procedure is also used in aircraft design. Here, the aim is to make precise predictions of the wing vibration in order to avoid the wing vibrating so hard that it breaks off. Many other effects in airplane design can be studied using the finite element method, such as the acoustic radiation of the engines on the cabin. Such a calculation has been done – with software of the Flemish-German company LMS-Siemens – for a sport plane (see Fig. 9.2). Figure 9.3 shows the division of an aircraft into finite elements. Based on the finite element calculations it can then be verified whether the design meets all requirements.

An additional advantage of finite element calculations is that improvements can be tested without having to build a real prototype just to give it a

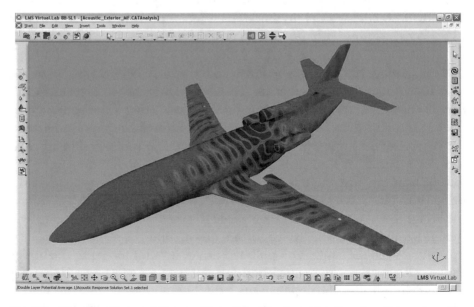

Fig. 9.2 The vibration and the resulting distortion to which the aircraft is exposed by the noise of the engines. The red colour indicates the areas that undergo the most distortion

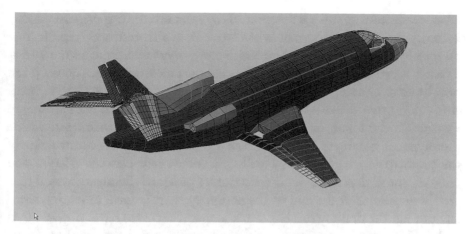

Fig. 9.3   An aircraft is divided into finite elements to calculate in-flight vibrations

try. Still, virtual experiments with finite element models come with their own problems, such as the computational effort that is required to obtain their solution. For the Boeing 787 Dreamliner, for instance, the finite element model that was built was so detailed that an ordinary desktop computer would take about 92 years to calculate the solution. Fortunately, Boeing can use a very powerful Cray supercomputer for those calculations, which is thousands of times faster than your desktop at home. Still, even on supercomputers, such computations can take days, rather than minutes.

Do the calculated vibrations actually occur in the plane? Of course, it is crucial that the mathematical finite element model approaches reality well, something that is not always self-evident. That is why – when we have confidence in a design – a prototype is still built, on which very extensive tests are carried out before the aircraft is allowed to take to the air. Trusting computations blindly without real-life testing is something even the most optimistic mathematical engineer would not recommend. These tests are called ground vibration tests and are used to validate and improve the mathematical model used. To carry them out, a device that subjects the aircraft structure to certain forces is attached to the wings at various places. The resulting vibrations are then measured and compared with the vibrations in the finite element model. By alternating physical and virtual prototypes, the number of physical prototypes can be greatly reduced. For the Boeing 787, for example, only 11 prototypes had to be made for the wings, while for the Boeing 767 there were 77 prototypes.

Also crash testing of cars replaced by virtual testing Virtual testing becomes even more important when we look at the design of cars. In that case, a physical test means that we crash a prototype of a car against a wall at a certain speed and film it in detail. Because the prototype breaks down, it can only be used for a single test. This makes physical crash tests extremely expensive. Add to that the cost of the dummies in the car, packed with sensors to measure all the forces on the driver and passengers. A physical test with a prototype can cost up to 100,000 euros, without counting the construction of the prototype itself. A dummy will cost somewhere between 50,000 and 600,000 euros, depending on the electronics provided. Because physical tests are always needed at the end of the design cycle, they have to be purchased anyway, but the dummy can last much longer if there are fewer physical tests with it. For complex products, a recent study shows that the number of prototypes needed decreases on average from 4.6 to 3 thanks to the intensive use of virtual testing. With an average cost of 1.2 million euros and a construction time of 100 days per prototype, this reduces the average development time for a new product by 158 days and the average cost by about 1.9 million euros.

Guaranteeing the safety of new football stadiums Virtual prototypes are also used routinely in the design of large structures, such as football stadiums. This type of construction has to meet a lot of requirements. The safety of spectators must be guaranteed when entering and leaving the stadium, as well as during emergency evacuations. There must be no traffic jam around the stadium and access to the stadium must be efficient. Many standards have been drawn up for this purpose, both by local authorities and by organisations such as UEFA and FIFA. In these circumstances, physical tests on prototypes are no longer possible at all! We can't simply build a complete stadium just to establish afterwards that it does not meet the safety standards. A scale model is not an option either, because we cannot shrink the spectators. All that remains is simulation, virtual testing. Fortunately, that's possible. Many experimental studies have already been carried out that show how people move, depending on numerous factors, such as the surrounding crowd and the alcohol level in their blood. It is only a matter of describing the results of those studies in a mathematical model for the movement of football spectators and placing them in a virtual realisation of the football stadium to be built. There are several companies active in this sector, such as the Dutch CrowdProfessionals. These kinds of calculations are nowadays standard for many public places, such as shopping malls, concert halls or festival grounds.

Virtual prototypes are not only important for cars, bridges and airplanes but also for lawn mowers, vacuum cleaners, washing machines and so on. For just about every product that comes on the market, the quality, behaviour and usability is tested on a prototype by the manufacturer before mass production. This involves not only vibrations and collisions but also, for example, the electrical operation of control circuits, the magnetic radiation of mobile phones, the stability of drones or the strength of satellites launched into space. Even the impact of new roads or intersections on the local traffic situation is simulated before the works can actually start. This makes it possible to make a reasoned estimate of complex designs without having to create the physical object. Now that production of consumer goods is quickly moving to countries with lower wages, our economic added value must therefore come from technological innovation that allows reducing costs. Virtual prototypes are an excellent tool for this.

Chapter 10
The Art of Forgery

'Pics or it didn't happen!' It is the modern version of the old 'Seeing is believing' and the phrase is often used on the internet when someone is clearly bragging. But can we just believe what we see? Can we check the authenticity of a photo? And what does that question have to do with the Lamb of God?

What's the problem? A photo can be faked in many ways. In advertising, it is commonplace to edit the models into the perfect beauty ideal. Red Hot Chili Pepper John Frusciante was digitally removed from a photo for Rolling Stone magazine in 1992 because he had left the band in the meantime. In the 1930s, Stalin did something similar with Trotsky (and that without a computer!). And in 2010, BP manipulated dozens of photos after the oil spill in the Gulf of Mexico to exaggerate their efforts in limiting environmental damage. Can we detect these kinds of modifications? Can we protect photographs from forgery so that they can serve as legal evidence? And while we are at it, can we also detect forged paintings?

The mathematical language of an image To better understand a digital image, we must first make a mathematical representation of it so that we can perform calculations. Fortunately, this is not so difficult: an image is essentially nothing more than a rectangle, divided into pixels, with each pixel having a precise colour. If we link a numerical value to each colour, we obtain a matrix that contains exactly the same information as the image. For example, Fig. 10.1 shows a simple drawing of a house. You can clearly see grey and black pixels. The other pixels are white. We choose to represent white as a 0, grey as a 1 and black as a 2. Next to the drawing you see a piece of the matrix representation, especially the piece in the middle at the bottom with the door in it.

G. Samaey, J. P. L. Vandewalle, *The Invisible Power of Mathematics*, Copernicus Books, https://doi.org/10.1007/978-1-0716-2776-1_10

$$\begin{bmatrix} 0 & 0 & 0 & 0 & 0 & 0 & 0 & 0 & 0 & 0 \\ 0 & 0 & 0 & 0 & 0 & 0 & 0 & 0 & 0 & 0 \\ 0 & 0 & 0 & 0 & 0 & 0 & 0 & 0 & 0 & 0 \\ 0 & 0 & 0 & 0 & 0 & 0 & 0 & 0 & 0 & 0 \\ 0 & 0 & 0 & 1 & 1 & 1 & 1 & 1 & 0 & 0 \\ 0 & 0 & 0 & 1 & 0 & 1 & 0 & 1 & 0 & 0 \\ 0 & 0 & 0 & 1 & 1 & 1 & 1 & 1 & 0 & 0 \\ 0 & 0 & 0 & 1 & 0 & 1 & 0 & 1 & 0 & 0 \\ 0 & 0 & 0 & 1 & 1 & 1 & 1 & 1 & 0 & 0 \\ 0 & 0 & 0 & 1 & 1 & 1 & 1 & 1 & 0 & 0 \end{bmatrix}$$

Fig. 10.1 Left: Drawing of a small house with three grey tones: white has value 0, grey is 1 and black is 2. Right: Matrix representation of a piece of the drawing, specifically the piece in the middle at the bottom with the door

In general, three matrices are kept for an image, one with the red, green, and blue values, each of which can vary between 0 and 255. All possible combinations then represent $256^3 = 16\ 777\ 216$ colours. We can then look for patterns in the matrix representation of the image. Do the same blocks of numbers occur more than once? Then probably a piece of the image has been copied. Do the values of some numbers match those of their immediate neighbours? Then probably something was pasted into the image.

Digital watermarks Now that we can calculate with images, we can suddenly do much more than just discover tampering. For example, we can change pixels ourselves – in a controlled way – to protect an image from manipulation. This can be important to allow the use of photographs as evidence in a court of law but also to safeguard the photographer's copyrights. An important technique is digital watermarking. In its simplest form, a watermark is applied to a photo by hiding a second photo in it using mathematical operations. If the image has not been tampered with, we can reverse the calculations to make the watermark appear again.

Digital watermarks are often used to determine the origin of leaked images. They are placed on films that are handed out to judges for the Oscars to prevent the film from circulating illegally on the internet before it is in theatres. There is also a digital watermark in every digital film in the cinema that contains information about the auditorium and the time of the screening. When a viewer shoots the film from the auditorium, it is possible to discover which cinema is not taking security seriously.

Which Van Gogh is the forgery? Since the early 2000s, digital image processing has also been used to check the authenticity of paintings. To do this, the paintings are photographed at a very high resolution, and subjected to a statistical analysis: a number of characteristics are calculated for each part of the painting, such as the density and length of the brush strokes. The style of the painting is thus represented by a limited number of characteristics, typically a few dozen. For each painting, these numbers are then turned into a vector. When the calculated characteristics are chosen in such a way that they form a good measure to quantify a painter's style, two vectors appear to be close together when they belong to two works by the same painter.

Starting in 2007, a group of researchers, led by Rick Johnson (Cornell University), made this effort for the works of Vincent Van Gogh, in collaboration with the Van Gogh Museum in Amsterdam. Dozens of works were once wrongly attributed to Van Gogh. Some by intentional forgery, others by coincidence in the legacy of Vincent's brother Theo Van Gogh. (Vincent's brother Theo should not be confused with the film director of the same name who was murdered in 2004. The latter Theo is the great-grandson of the first.) A well-known example of deliberate forgery is the Wacker scandal that broke out in 1928, in which the German art dealer Otto Wacker had 33 (fake) 'early Van Goghs' auctioned off 'from the collection of a rich Russian who wished to remain anonymous'. The fraud only came to light years later. Such a big collection of originals and excellent forgeries was a dreamed object of study for the researchers. Van Gogh's style was quantified on the basis of 101 paintings (both genuine and forged). The results are still being used to this day to verify the authenticity of Van Gogh's paintings from private collections offered for auction.

Wavelets and the Belgian connection

Over time, cracks appear in the paint layers of paintings due to aging. To detect these cracks, we can use mathematical image analysis. For mathematicians, an image is no more than a rectangle divided into boxes (pixels), with a number in each box representing the colour of that pixel. In a picture of a painting, the cracks in the paint correspond to places where we see a sudden colour transition between neighbouring boxes. A wavelet analysis calculates a different representation of the same image that allows these sudden transitions to be read off directly.

For convenience, we explain the principle for a 1-dimensional 'image', as shown left on Fig. 10.2. The curve represents a line in the image that contains two colours (represented by the numbers 10 and 5) and two very sharp cracks (represented by a 0). The line contains 128 pixels.

(continued)

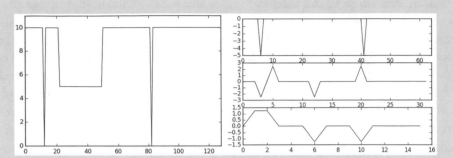

Fig. 10.2 Left: A single line of 128 pixels from a photo. The line consists of two colours (with values 10 and 5, respectively) and two small cracks (represented by a 0). Right: From top to bottom the 3 finest rows of the wavelet decomposition. The top row shows only the sharpest transitions of the cracks, the middle row also shows the colour transitions. The lower row is more difficult to interpret by sight

We also limit ourselves to a very simple wavelet (known by specialists as the Haar wavelet). In that wavelet representation, we calculate two new rows, each half as long (i.e. 64 pixels each), from the original row of numbers (the figure on the left). The first row contains the average value of two consecutive pixels. The second row contains the deviation of these two individual from that average. (The right pixel always deviates from the average as much as the left pixel, but in the other direction). Those two shorter rows together contain exactly the same amount of information as the first longer row. So the transformation is reversible and the original image can be restored.

Now we move on. First we keep the second row. (That is the row with the differences.) We show it at the top right of Fig. 10.2. In that figure, we can already clearly detect the cracks. Then we continue with the first row (with the averages) and repeat the operation. So again we calculate two rows (now each of length 32). The first row now contains the average of four consecutive pixels. The second row then contains the difference between the average of the two on the left and the average of the two on the right. Again, we keep track of those differences (middle right on Fig. 10.2). In our example, they represent transitions between slightly larger areas of the same colour. Here, too, the cracks remain visible. We can continue this way. In the end, we still have two values left: the average colour of the whole image and the extent to which the left and right half deviate from that average.

You might wonder what's the point of this: we can see those cracks with the naked eye, can't we? That's right, but it's very difficult for the human eye to pretend they are not there. With wavelet analysis it suddenly becomes very easy to make an image with *only* those cracks and then subtract that image from the original – a digital restoration, as it were.

Chauvinistic as we are, we cannot fail to mention that with Ingrid Daubechies and Wim Sweldens some Belgians were at the cradle of this success story.

The restoration of the Ghent Altarpiece (Adoration of the Mystic Lamb) by the brothers Van Eyck We don't need to stop at the discovery of forgeries of paintings: mathematical painting analysis is also used for the study of damaged paintings. This may involve, for example, uncovering overpaintings or discovering which pieces of painting are the result of previous restorations and which pieces are original. In the case of the Adoration of the Mystic Lamb by the Van Eyck brothers, for example, craquelures have arisen in the layers of paint over the years. In the panel on the Annunciation of Mary, the painted text has become illegible as a result. For centuries, art historians have wondered whether this text is based on an existing manuscript. Led by Ann Dooms (Vrije Universiteit Brussel), a team of researchers succeeded in digitally removing the craquelure pattern using wavelets on a picture of this panel with a resolution of 100 billion pixels. Obtaining that picture in itself was a feat, because a camera with such a resolution obviously does not exist. The photo is therefore a digital collage of photos of each of the parts of the painting. The researchers noticed that the text seemed to correspond to a manuscript by Thomas de Aquino (1225–1274), more specifically with an adaptation that dates back to 1430. Not unrealistic, given that the Adoration of the Mystic Lamb was painted in 1432. The methods are now eagerly used as an aid in the large-scale restoration of this painting, which started in 2012 (Fig. 10.3).

Fig. 10.3 Digital restoration of a detail from the Mystic Lamb. Left: painting with cracks. Middle: Reconstruction of only the cracks, obtained through a wavelet decomposition. Right: Digital restoration in which the cracks have been digitally removed. The text is now clearly readable B. Cornelis, T. Ružić, E. Gezels, A. Dooms, A. Pižurica, L. Platiša, J. Cornelis, M. Martens, M. De Mey, I. Daubechies, "Crack detection and inpainting for virtual restoration of paintings: The case of the Ghent Altarpiece," Signal Processing, Volume 93, Issue 3, 2013, Pages 605-619, ISSN 0165-168

The importance of the protection and security of images can hardly be overestimated. Just think of the economic value of copyrights, the legal evidential value of a photograph, the cultural value of works of art or the political power of a provocative image. Every day, digital image analysis shows the possibilities that arise when mathematicians and specialists in each of the mentioned domains join forces.

Chapter 11
The Right Bike in the Right Place

As an alternative to the car or public transport, many cities are investing in systems that allow users to borrow a bicycle at one point and return it somewhere else. By June 2014, 712 cities around the world had such a system for bicycle sharing, totalling roughly 806,200 bicycles and 37,500 bicycle stations. But how are these bicycles distributed among the stations? And how can you be sure there will be a free space to park your bike at your destination?

What's the problem? When people can borrow a bicycle at a place of their choice and drop it off at another place, imbalances naturally arise in the distribution of bicycles among the bicycle stations. People prefer cycling downhill to uphill. They cycle from the railway station to work in the morning and they only return in the evening. Tourists cycle to popular attractions and leave their bikes behind upon arrival. People cycle more often and longer when the sun shines than when it rains. And so on. To make sure all users are well served, small vans are constantly driving around to properly redistribute the bicycles. Such a van typically has room for about 20 bicycles. In the Belgian city of Antwerp alone, these journeys account for some 75,000 km on an annual basis (at least in 2013, then for some 1000 bicycles). To reduce costs and maintain the environmental benefits of cycling, this redistribution should be done as efficiently as possible.

Divide and rule The redistribution of bicycles consists of two clearly distinct subproblems. First, we need to determine how many bicycles we *want to* have available at each station at any given time to meet the demand. For this, we have to take into account the weather, the day of the week, events, and so on.

83

G. Samaey, J. P. L. Vandewalle, *The Invisible Power of Mathematics*, Copernicus Books, https://doi.org/10.1007/978-1-0716-2776-1_11

Then we need to determine how we get the desired number of bikes in each of the bike stations: we have to determine the order in which we drive past the bike stations and how many bikes we have to load or unload at each stop. Moreover, there are a number of limitations: there is a maximum number of bicycles that fit in the van, the drivers have driving and resting times, and we would like to take the traffic situation into account. The resulting routing problem is extremely difficult to solve, even with the fastest computers.

When do we call a problem 'difficult'?   What exactly do we mean when we say that a problem is 'particularly difficult' to solve? The answer to that question comes from *complexity theory*, a branch of mathematics that studies how much time it takes computers to solve problems, depending on the type and size of the problem. For bicycle sharing, the relevant problem size is the number of bicycle stations. Of course, more time is needed to solve larger problems, but for certain types of problems the calculation time rises faster as a function of the problem size than for the other. Take, for example, the search for the largest number in a row of n numbers. The calculation time needed for that problem is proportional to the length of the row of n, because all you have to do is walk through the row once and remember the largest number on the way. Sorting a row is a bit more difficult: if you do so in a naive way, you might decide to first search the largest number in the row and put it at the end of the row. When that is done, you know the last number is at its destination, and you repeat the same procedure for a slightly shorter row. You can keep doing this until all numbers are sorted. The amount of work now becomes proportional to n^2, because you need to look n times for the largest number in rows of length $n, n - 1,..., 2, 1$. (There are sorting methods that are much faster, but that is not the point here). You notice that the calculation time for this algorithm increases by a factor of 4 when the length of the row to be sorted doubles (because $(2n)^2 = 2^2 \cdot n^2 = 4n^2$).

Exponential complexity   Both problems above are relatively simple: the necessary calculation time rises as a power of the problem size. The higher that power, the faster the necessary calculation time increases with the problem size. When searching for the largest number, doubling the length of the row doubled the calculation time. When sorting the row, this became a factor of 4. The important thing to note is that there is a constant factor by which the amount of computations grows when the size of the problem doubles. In the case of problems that we call particularly difficult, such as the routing problem for bicycle sharing, the calculation time increases much faster as a function of the problem size. To imagine this, it helps to think of an exponential relationship between problem size n and calculation time, like for example 2^n.

Note that the problem size itself is in the exponent now. Such a complexity is very problematic. Imagine that we increase the problem size by one. For the problem of bike sharing, this simply means adding one bike station. With the exponential calculation time that we postulated, this addition immediately doubles the necessary calculation time. Simply observe that $2^{n+1} = 2 \times 2^n$. And we haven't even doubled the number of bicycle stations, we just added one! Suppose we actually double the number of bicycle stations, then the calculation cost for the larger problem suddenly equals the square of the calculation cost for the smaller problem, because $2^{2n} = (2^n)^2$. The bigger the problem, the more problematic a doubling of the problem size becomes! To give an idea for bike sharing: if routing with 10 bike stations in the above case would result in a calculation cost of 2^{10} (about 1000), we would arrive at 2^{20} (about 1,000,000) for 20 stations. Take 40 stations, and we arrive at a cost of $(1,000,000)^2$, or 1000 billion. It is clear that we are not going to get anywhere for problems of a decent size.

The P versus NP problem

Problems come in different shapes and sizes. For convenience, we stick labels on them. Problems in category 'P' are problems for which a 'fast' solution method exists: the necessary calculation time for a fast solution method is proportional to a power of the number of unknowns. For problems in category 'NP' we can 'quickly' check the correctness of proposed solutions. Checking a solution is often easier than finding a solution: for example, it requires less work to see if a row is sorted than to actually sort a row.

We know the bicycle sharing problem is in the category 'NP': here checking the optimality of a solution is relatively simple, but finding the optimal solution is very difficult – and this purely because there are so many possible solutions. Of course all 'P' problems are also 'NP'. But is the converse also true? If you look at the bicycle sharing problem, you would think it isn't. We only have at our disposal 'slow' solution methods for the bicycle sharing problem that have an exponential complexity. But that does not mean that no fast solution method exist! It just means that we have not yet discovered any fast method yet!

The 'P versus NP problem' is exactly that question: can a problem always be solved 'quickly' if one can quickly verify that a proposed solution is correct? The P versus NP problem is unsolved, although the question was asked in a very precise mathematical form back in 1971. The person who comes up with a proof (whether the answer is 'yes' or 'no') can be rewarded with a *Millennium Prize* of $1,000,000 from the *Clay Mathematics Institute*.

It has to be said that it is not entirely clear to everyone what the practical impact of the evidence would be: if the answer is 'yes', we don't necessarily have a good algorithm right away, just the certainty that an algorithm must exist. Moreover, it might be safe to assume that the algorithm is theoretically in the 'P' class, but that the (constant) speed at which work grows when doubling problem size is so big that for large problems it is still infeasible to calculate the exact solution with that 'fast' method.

Reasonable is good enough The major stumbling block with routing problems is that there's really not much else one can do than proposing solutions and then check how efficient they are. It can help a bit to think about the order in which we look at possible solutions – with a bit of luck we can discard whole groups of solutions without the explicit need for a detailed check – but in essence we just go down the road until we have the most efficient solution. As is often the case in life, if we were just satisfied a little more quickly with a reasonable solution, we could save ourselves a lot of trouble. This brings us into the realm of *heuristic* methods: methods that use tricks to get an acceptable solution quickly, with no guarantee that the solution is really optimal. Of these methods, genetic algorithms appeal most to the imagination.

Survival of the fittest Genetic algorithms propose a whole stack of solutions as a starting point. For bicycle sharing, these solutions are ride schedules for each of the vans. Then, following the Darwinian model of evolution, we introduce two mechanisms to generate new solutions. The first is 'genetic recombination' – sexual reproduction as it were. The rules for genetic recombination are used to create a next generation of solutions (the children) that combines specific characteristics of two existing solutions (the parents). In the case of bicycle sharing, this means that we compose a new generation of ride schedules by randomly combining pieces of rides from each of the two solutions in a consistent way.

The second mechanism is 'mutation', or random change. In this phase 'genetic mistakes' are made during reproduction. In the Darwinian model of biological evolution, mutation is the mechanism by which species evolve and become better adapted to their environment by chance. In the genetic algorithm, mutation is a way of adding pieces of rides that were not in the original set of rides.

In the Darwinian model, children who are better adapted to their environment have an evolutionary advantage. They live longer and have a greater chance to reproduce. In genetic algorithms, this 'adaptation' is entered as a score that measures the quality of a solution. We therefore calculate a score for each of the proposed ride schedules that indicates how 'good' it is. The solutions with the lowest scores then 'die'. These can be parents as well as children. With the better scoring solutions we 'breed' further. We stop as soon as an acceptable solution is calculated.

The principle of genetic algorithms is universal. By combining relevant factors generation by generation, we efficiently seek a solution that achieves the set standards. The difficulty lies, of course, in determining the precise rules for reproduction and for determining quality. This currently requires a great deal of experience and specific insight into the problem.

What about multiple goals? The routing problem is actually even more complicated than what was described above. To start, it is not entirely clear what we mean by the most efficient solution. We can have several goals at the same time. For example, it can be interesting to minimise the distance travelled by the trucks, and the number of times the drivers have to load or unload. It is not clear upfront how we should weigh these two different types of targets against each other. That is why we look at the different goals side by side for each solution and look for solutions that do at least as well as other solutions for all goals. That is what we call 'decent' solutions.

Is any solution feasible? In addition, it is sometimes not possible to meet all conditions at the same time. For example, a driver may not be able to make his rounds quickly enough if he has to comply with all the traffic rules at the same time. Then the algorithm must be able to choose which restrictions will be ignored and which ones will be enforced. (In this case, preferably the traffic rules.) So we have to distinguish between conditions that are strictly necessary on the one hand, and preferences and wishes on the other hand. Typically, this is done by specifying necessary conditions as preconditions: a solution will not be considered if it does not meet those conditions. Preferences are considered as costs: we can consider solutions that ignore the preferences, but any ignored preference is at the expense of lowering the quality score of the solution.

The current state of the art Currently, most redistribution of bicycles is done manually, with a central person monitoring how many bicycles are in each station. For cities like Vienna, with about 1500 bicycles, this is a full-time job. This is an important obstacle to increasing capacity. In 2014 a pilot project was started in Vienna in which the central person could use a genetic algorithm for planning. The first results were positive, although there are also some problems. For example, the suggestions for the van drivers sometimes seem strange, especially at busy times. For example, they occasionally became very nervous when the system indicated that they should take fewer bicycles from an overfull station than what they could fit in their van. (An algorithm could suggest that, for instance, because the situation at the next bicycle station might even be worse, and another van is close by anyway.) Some drivers even ignored seemingly illogical suggestions, which of course completely confused the algorithm. In Antwerp, too, experiments are currently being conducted with computer help for the redistribution of shared bikes.

The social importance of planning cannot be overestimated. Not only bicycle sharing but all logistics have to deal with it, from delivering products to shops or at home, to making timetables for postmen and postwomen to planning public transport. Even schedules for prisoner transport are drawn up using such optimization methods. These methods are also needed for other planning problems, such as timetables of staff, for example nurses. Because for all such complex problems, the optimal solution requires an unreasonable investment in computation time, it is usually decided in advance how much calculation time the computer gets, and one proceeds with the best solution that will emerge within that time. Any method to make the calculation process more efficient thus leads to better solutions within the same calculation time and is therefore extremely valuable. After all, a better solution means economic profit and less frustration. What else can a mathematician dream of?

Robert Bridson, Mathematician and Oscar Winner

If you've recently watched a Hollywood blockbuster, you've almost certainly come into contact with the work of Robert Bridson, a Canadian mathematician involved in computer animation. In 2014 Bridson received a technical Oscar for the mathematical methods he developed to create film scenes on computer. His work provided the visual delights that contributed to the success of top films like *The Hobbit: The Desolation of Smaug, The Adventures of Tintin, Avatar, Life of Pi* and *Gravity*.

Bridson predicts what a certain event will look like using computer simulation with mathematical models. But while such simulations usually serve to make precise scientific predictions, here the artistic representation of the scene is especially important. For example, it is not unusual for a director to be dissatisfied with a realistically simulated fire and ask for the plume of smoke to be made 'a little more threatening'.

In the early days, computer animation was, of course, mainly used to create scenes that are difficult to record in a different way. In *Gravity*, for example, a fire starts at some point in the spaceship. For the filmmaker it is important that such a fire looks realistic, but it is a fire in an environment with no gravity. How do you film something like that? Simple: by simulating the equations for combustion on the computer and turning off the effects of gravity in that simulation!

Even when the images are technically feasible, they are often generated by computer for artistic reasons. Like a fragment in *Avatar* on which a number of 'creatures' are already flying between ferocious sea waves crashing into rocks. These 'creatures' of course do not exist. So they are the result of computer animation. But that's not the only thing that's fake in that scene. From the clouds in the sky to the waves and the rocks, the whole scene is generated on the computer. The reason for that is surprising. If you pay close attention, you will notice that the impact of the waves on the rocks is perfectly synchronized with the film music. Real waves can't be directed that way, so the filmmakers grabbed a piece of software.

Meanwhile, computer animation has become a convenient and inexpensive way to avoid risks. Suppose you want to film a conversation between two actors on a street in New York, then that's technically possible. But of course you need a license to film, you have to clear the street, and fly in your material and actors to New York. And then, when everything is sorted out, it's a rainy day and the script asks for sun. The cost of such a setback quickly rises to several hundred thousand euros. Today, the visual effects are so advanced that such scenes can just be recorded in a studio. The street in New York is glued in afterwards, no one notices.

Bridson made his crucial contributions in his Ph.D, which he defended at Stanford in 2003, and made them available through the Naiad software package. Without him, our films would look a lot less spectacular.

Chapter 12
Can Computers Detect Fraud? (And Do We Want Them To?)

Our brains can do a lot. We recognise others. We read license plates, road signs and signposts while driving, almost effortlessly. Doctors make routine diagnoses, even for very rare diseases. Financial experts recognise fraud 'when they see it'. Psychologists are trained to quickly assess someone's personality. Can a computer do all that? And if so, do we want that?

What's the problem? The examples above have something in common. None of them is a gift that we are born with. They all represent a skill that some of us have *learned*. Each example is about pattern recognition, whether it's numbers or faces, illnesses or financial transactions. Often we don't know exactly *how* we recognise something, nor can we indicate exactly *why* something is beautiful or suspicious. We just *feel* it, the pattern recognition is so firmly wired in our brain that it has become an automatism. By the way, it might be interesting to point out that our brain is evolutionarily trained so well for pattern recognition that we often even see connections that don't exist at all. The work of the American psychologist Kahneman contains numerous examples of this phenomenon.

There is also a case for using computers for pattern recognition. Owing to advances in technology, so much data is collected and maintained that it is simply infeasible to have everything processed by humans. Think, for example, of isolating a few suspicious financial transactions. Moreover, there are many situations in which the available data is not suitable for human interpretation. Genetic information is a typical example. And computers are probably more immune to human errors of thought or randomness when detecting patterns. Can we make the computer as efficient as the human brain for these kinds of tasks, or maybe even more efficient? If so, how? And what are the benefits and dangers?

© The Author(s), under exclusive license to Springer Science+Business Media, LLC, part of Springer Nature 2022
G. Samaey, J. P. L. Vandewalle, *The Invisible Power of Mathematics*, Copernicus Books, https://doi.org/10.1007/978-1-0716-2776-1_12

The distinction between good and bad Let's start with a clear black and white decision: a doctor has to determine whether a patient has a certain disease or not. He has a number of parameters at his disposal: blood values, medical images, and the clinical story. Unfortunately, there are not always simple rules to determine a correct diagnosis on the basis of these data. The doctor has a great deal of theoretical knowledge of the diseases and has made a considerable number of diagnoses in the past, either positive or negative. In order to come to his conclusion with this new patient, he tests that patient against all the available information. However, a doctor is a human being and sometimes makes mistakes. He may overlook things, be tired or stressed, or suddenly be confronted with a clinical picture with which he has no experience. A computer, on the other hand, is never tired and has direct access to all the information that has ever been put into it. Can a computer help the doctor?

To do so, the computer must have a way of classifying a new patient as 'healthy' or 'sick', based on the data of patients whose diagnosis was made earlier. Mathematically, it is a matter of converting all of the patient's characteristics into numbers. There may well be dozens of them. For simplicity's sake, suppose there are 30, but that number is just an example. Those 30 numbers represent the data of one person as a point in a 30-dimensional space. Because we cannot draw in 30 dimensions, we pretend for a moment that we have only two characteristics: they then represent points in a plane, like in Fig. 12.1. So for each of the former patients, we pretend to have two measurements. We also know for each of these patients whether they were

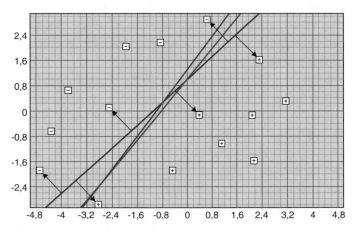

Fig. 12.1 Measured values of patients in a plane. The horizontal axis and the vertical axis show possible measurement values of two quantities. Each study of a patient results in one value for each of the two quantities, i.e. one point. '+' points belong to sick patients, '−' points belong to healthy patients. The three straight lines separate healthy from sick patients. Of those lines, the red one is the 'best'

'sick' or 'healthy'. We indicate this on the figure by marking the sick patients with '+' and the healthy with '−', as is customary for diagnoses in medicine: a 'positive' diagnosis is usually bad news. The problem then becomes: can we divide the plane into two regions, one region for the sick patients and another region for the healthy patients?

Very often that separation can happen with a straight line. That is the case, for example, in Fig. 12.1. When we now have to assess a new patient, it suffices to look along which side of the line that the measurements for that patient fall to make the diagnosis.

However, there are still a number of practical problems. First of all, there are several straight lines that separate the '+' dots from the '−' ones. In Fig. 12.1, the red, the blue and the green lines are all correct separations. Which separating line we choose makes a big difference for 'borderline cases': it can depend on our choice along which side they fall whether they are considered positive or negative. We therefore choose the line that is furthest from all our measurements so that we minimise the chance of edge cases. In our example, this is the red line. The other lines pass rather closely to some of the measuring points. Of course, the best thing to do is to subject the edge cases to extra examinations and a more detailed look by the doctor...

A second practical problem is that the '+' and '−' points can often not be separated by a straight line, but that a curve is needed, as shown in Fig. 12.2.

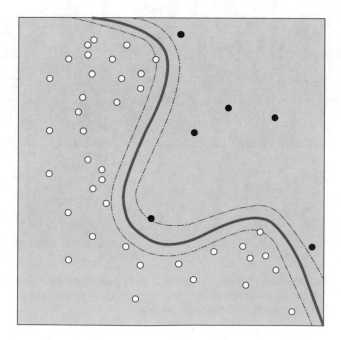

Fig. 12.2 The measured values of sick patients ('+' dots) and healthy ('−' dots) can only be separated here by a curve

If we allow curves, we can be much more flexible in solving classification problems, but finding the 'best curve' becomes a lot more difficult. In this case we call the problem *nonlinear*.

Isolating suspect cases   There are many other problems where we may want to look for a curve to separate two types of cases. An important example is fraud detection when using credit cards. Again, it is not possible to describe in simple rules when we should consider the transactions as fraudulent. Credit card companies do have an extensive collection of past transactions of users and also a fairly large number of examples of past fraud cases. This is not enough to label a transaction as fraud with certainty, but it is enough to be able to quickly consider the vast majority of transactions as 'normal'. Out of the millions of transactions per day, possibly fraudulent cases can then be isolated by computer, after which only the suspicious transactions need to be examined in detail by the experts of the credit card company. It is very important not to unnecessarily block a good customer's card. It turns out that we typically need a *closed* curve in such cases. In Fig. 12.3 it is presented as an ellipse, but in general determining the shape of that curve is part of the problem. The normal transactions are now inside the ellipse, the suspect outside it.

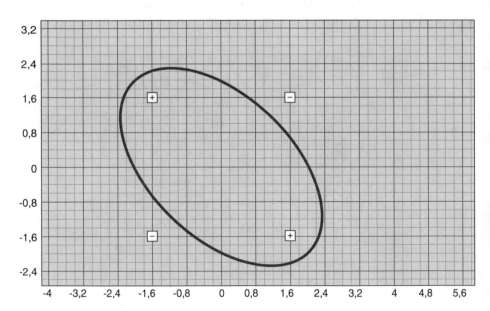

Fig. 12.3   In fraud detection, fraud cases are usually separated from normal transactions by a closed curve. Measurements that fall outside the 'normal' area are then suspect and are further inspected by fraud experts

Looking for the right curve For our classification problems, we generally have to look for a curve (open or closed) that separates two types of points in a high-dimensional space. Because the shape of the curve is completely free, this problem is much more complex than separating using a straight line. To explain exactly where the greatest difficulty lies, we imagine again that each measured point – in this case: every registered transaction with a credit card – is represented by roughly 20 parameters. These include, for example, the place and time of the transaction, but also the cardholder's place of residence and, for example, their travel behaviour. (A sudden purchase in Shanghai, for example, is more suspicious if it happens on a card of someone who never travels abroad). The separation curve is then written as a nonlinear relation between those 20 parameters. Unfortunately, there are so many nonlinear possible relations that it is almost impossible to start! So we have to try to simplify the problem – at least at first sight.

Artificial neural networks In the 1980s, a practical solution became available in the form of artificial neural networks. The idea is based on insights into how the human brain enables us to learn. Our brains consist of 100 billion neurons, which are interconnected by synapses. In Fig. 12.4, we represent a small number of neurons as spheres, and the synapses that connect them as lines. The neural network sends impulses from the input of the network through layers of neurons, via the synapses, to the output. The strength of the synapses can vary, and the importance that a neuron attaches to the information coming from the various other neurons connected to it depends on the

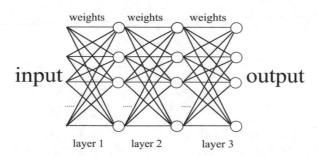

Fig. 12.4 Schematic representation of a neural network. Each of the spheres represents a neuron. Each of the lines is a synapse (a connection between neurons). Information flows through the network from left to right. Each column represents a layer of neurons that processes the input from the left and transmits it to the next layer of neurons on the right

strength of those connecting synapses. At the output comes the interpretation. In this way, for example, light that enters our eyes is converted to the text we read on a signpost along the road.

An artificial neural network is a computer program that mathematically attempts to mimic the learning character of our brains. This attempt is a strong simplification that contains only the essential properties of the brain. The aim is to recognise patterns the way that people do. To achieve this, we create a network composed of artificial neurons and synapses. These artificial neurons transmit impulses from the input to the output via the artificial synapses. The size of the impulse at the input depends on the incoming signal. Further on in the network, that size depends on the importance of those synapses. The neuron then calculates a certain nonlinear function of the weighted sum of the incoming pulses and transmits the result to the next set of neurons.

The mathematics behind artificial neural networks

You may not have noticed it yet, but with an artificial neural network, strictly speaking, we do not exactly solve the same problem as the one we started from. The difference is subtle. Our original question was: find a nonlinear function of the 20 magnitudes that distinguishes two cases. However, the artificial neural network tries to find this function in a very specific form: each neuron makes a weighted sum of the incoming signals, and then calculates a nonlinear function of the result. Is this subtle difference important? Or are both questions identical? The latter would be the case if any nonlinear function of the 20 unknowns can be obtained as a solution of a neural network. That this is the case, is not obvious.

The question has occupied mathematicians throughout the twentieth century. It started with the German David Hilbert, one of the leading mathematicians at the beginning of the twentieth century. Hilbert not only did pioneering work himself, he also asked a lot of questions that were to serve as inspiration for others. In the year 1900, he formulated his 23 problems for the twentieth century. The question above is known as problem number 13. Note: no one was thinking about artificial neural networks at that time. Hilbert asked the question purely out of theoretical interest. He also thought that the answer would be negative. According to him, sooner or later an example would be found of a nonlinear function that could not be decomposed in the way an artificial neural network does.

The answer was to wait until 1957, when two famous Russian mathematicians (Kolmogorov and Arnold) proved mathematically that the above two questions are indeed equivalent. Hilbert had already died 14 years earlier. Unfortunately, their proof was not constructive: Kolmogorov and Arnold did not provide a practical way to exploit that equivalence. In order for that one had to wait for the breakthrough of artificial neural networks in the 1980s.

Artificial neural networks must be trained   The analogy with the human brain goes even further. Just as our brains need to learn, artificial neural networks also need to learn the difference between normal payments and fraudulent ones. For example, when children learn to recognise animals, they will make mistakes in the beginning. Some children might just gamble. They will then be corrected by parents or teachers, so they won't make the same mistakes next time. Maybe we show the same pictures of the animals several times, just to get the right answer. Well, artificial neural networks work in the same way. In the beginning, the artificial neural network does nothing more than gamble. During the learning process, the data of a large number of financial transactions are offered. (Or of patients, if we return to the example of the doctor.) We give the artificial network process the data for each of those transactions and let it decide whether a transaction is fraudulent or not. Then we give the answer. When the network is wrong, it will adjust itself in such a way that it will give the correct answer the next time for that transaction. This adjustment is done by changing the strength of the synapses, among other things. In this way, the artificial network gradually becomes better at recognising fraudulent transactions.

Towards a real artificial brain?   The functioning of the human brain is many times more complicated than that of an artificial neural network. So there are natural limits to the analogy. In the real brain, neurons are not points, but complex structures, and there are different types of neurons. The synapses are 'paths' that connect the neurons. While learning, the brain creates new pathways, or intensifies the existing pathways. Existing paths can also be pruned. This whole complex operation is represented in an artificial neural network by adjusting the weights on the synapses. This is a strong mathematical simplification – a model – that only represents the essence of this process. The analogy with the human brain therefore serves mainly as an inspiration and to explain the principle of the method comprehensively. Artificial neural networks are lightyears away from real artificial brains.

An exam for artificial neural networks?   Once the neural network has learned enough to correctly interpret the data, it can be applied to transactions it has not yet seen. If it also interprets those transactions correctly, the network is well trained. It goes without saying that we will not use the artificial neural network to detect fraud if we are not yet sure that it is well trained. So it will have to take an exam. That exam obviously needs to be corrected. To do that, we need to have the right answers ourselves. Therefore, it is important to withhold some of the data you have already collected so that you can test its learning process. Similar to when we teach children how to count during class but also ask other questions on the exam that the teacher already knows the answers to.

How far can we go? In financial transactions, the use of algorithms such as artificial neural networks has become common practice to select possible fraud cases for further investigation. In the medical sector, IBM, for example, is strongly committed to this kind of method for automating diagnoses by their supercomputer Watson (that's the same computer that won the American television game *Jeopardy! in* 2011). The first goal is the diagnosis of cancer. To this end, IBM works with the *Memorial Sloan Kettering Cancer Center*, which treats some 130,000 cancer patients annually. It processes the entire scientific literature. Through the acquisition of companies such as *Merge Healthcare* for $1 billion, IBM will also gain access to a huge database of medical images and test results to train Watson's algorithms. This is because *Merge Healthcare* itself owns a number of user rights to the images created with the scanners they sell to hospitals. Meanwhile, there is an app for the iPad in which doctors can upload their test results and get an overview of the complete scientific knowledge that applies to their specific patient within a minute from Watson.

What about psychology? All these developments sound spectacular and worrying at the same time. Surely there must be domains where computers and algorithms have no business at all? Like estimating someone's personality? Well, computers are getting better at that too. In the United States, a company, *eLoyalty*, invested 6 years of research time (and 50 million dollars) to develop software capable of classifying the personality of people who call into a call centre with a complaint. For this purpose, the company analysed factors such as speech patterns, vocabulary, sentence structure and talk rate of 600 million phone calls. As a result, it is able to classify each caller into one of six predefined personality types within minutes. Based on this, the caller is connected to an employee with a compatible personality. The result is, according to the company's own tests, a halving of the call time from 10 to 5 min on average and an increase in satisfaction from 47% to 90%. This application was made possible by starting from a quantifiable personality analysis method developed by NASA in the 1960s. It was originally administered to astronauts by trained psychologists (not computers) and was intended to ensure a good relationship among astronauts about to be launched together into space.

Artificial neural networks are a very powerful technique, with many possibilities. Many applications are embraced by mankind without hesitation because we only see advantages in them. Examples are the rapid detection of a few fraudulent transactions from millions of payments or smart cameras that automatically recognise license plates when crimes are committed. Others may have privacy concerns, such as profiling potential customers for targeted advertising, as is done with Gmail, or with automatic face recognition on Facebook or by Google. And in other cases, people may even be worried about their jobs. Ultimately, the power of these kinds of self-learning algorithms goes to the heart of humanity, with all its psychological and ethical consequences. What do we allow the algorithms to do? With which tasks do we entrust them? Where are the boundaries? These are all questions for which it certainly can't hurt to think a little deeper about the possibilities and risks of this technology.

Chapter 13
Industrial Fortune Tellers Predict Profit

A good prediction is worth its weight in gold. Newspapers predict their sales to determine their circulation. Demographers predict birth rates to advise the government on school building. Electric power plants predict consumption to steer production. And most of us also find accurate weather forecasting very important. But forecasting means looking into the future. How does that happen? And how can we estimate the quality of forecasts?

What's the problem? Predicting the future is very important for many companies, especially when they take actions whose consequences only become clear after some time. In the steel industry, for example, blast furnaces serve to make liquid steel from iron ore and carbon. The temperature in the blast furnace is very important here. Too cold and the contents solidify, too hot and safety risks arise. The temperature is kept under control by adding (for example) air, scrap or cokes. However, such interventions only have an effect 6–8 h later. So we have to be able to *predict* the temperature at that later moment since we cannot directly measure temperature at some future time. A similar problem applies to the electric power grid: conventional power stations often need at least a day to start up or shut down. So we need to know today how much electricity we will need tomorrow. Of course, we shouldn't gamble wildly: we know a lot of physics and chemistry, and we have kept decent records of what has happened in the past. How can we make the best use of all this knowledge for the predictions we need?

Mathematical models To allow predictions, we need to use as much available information as possible. We do this by creating a *mathematical model* that reproduces everything we already know. A mathematical model is a collection

G. Samaey, J. P. L. Vandewalle, *The Invisible Power of Mathematics*, Copernicus Books, https://doi.org/10.1007/978-1-0716-2776-1_13

of mathematical formulas. It contains quantities that are relevant for the description of the current state of the system, such as a temperature, current or voltage. You may think of physical quantities, but a relevant quantity may as well be a price, population size, and so on. The mathematical model then expresses how those quantities relate to each other and how they influence each other. So, mathematical models work like a very precise and unambiguous language: they *describe* in minute detail what we know about reality – or at least those aspects of reality that we consider relevant to our problem. The aim is then to use that description of what we know (the mathematical model) to find explanations for the things we see (the observations and measurements) or to make predictions. So mathematical models are not reality, they are a representation of reality. It is perhaps best to compare it with a painting, something the Belgian surrealist René Magritte did in his famous painting "La trahison des images" from 1929. The message in that painting (the fact that a pipe is not the same as a painting of a pipe) was not immediately understood by the public. We can make fun of that confusion now, but we have to realise that we need to make the same distinction between mathematical models and reality as between works of art and reality. The painting is a good representation for a pipe that you can use to learn what a pipe looks like. But you can't use the painting to learn how to fill a pipe. With mathematical models it's exactly the same: they are useful, but it would be nonsense to try to use them for something they weren't designed for.

Mathematical descriptions of physical principles A first way to obtain a mathematical model is to write down mathematically some universal principles: conservation of mass, conservation of energy, action and reaction, increase of entropy, and so on. When we write these principles down concretely in a specific context, we obtain, for example, Newton's laws of motion (when it comes to moving objects) or the Schrödinger equation in quantum physics. Einstein's formula $E = mc^2$ is also a mathematical description of the principle that mass and energy are equivalent. There are many, many models of this type. Of course, writing down a model for a specific problem requires

Lorenz and the butterfly effect

You would expect that making predictions is rather easy once you have a good mathematical model. Unfortunately, it isn't. In some cases, one can prove that good predictions are impossible to make, even with the best models available. In the early 1960s, as a young meteorologist at the Massachusetts Institute of Technology (MIT), Edward Lorenz investigated the use of computers for long-term weather forecasting. At that time, computers were not as fast yet as they are today ('only' a thousand times faster than people). So Lorenz had plenty of time to get a cup of coffee while the computer was calculating for him. When he came back, he made a shocking discovery!

Lorenz had drawn up a mathematical model with 12 variables (such as temperature and wind speed) with which he wanted the computer to predict the weather. That day, he resumed an earlier calculation that had stalled, good for 2 months of simulated weather. The result was shocking: the results were not remotely similar to his earlier calculations. His first reaction was that something had to be wrong with the computer. At that time, vacuum tubes were the most important building blocks for a computer and the vacuum tubes often broke down.

However, a check quickly showed that the computer was intact. So what was going on? The cause turned out to be as silly as it was surprising. Lorenz' computer calculated its results with a precision of six decimal places, but printed only three digits so that the results would fit neatly on one line. So when entering the initial state for his repeated simulation, Lorenz introduced a difference in the fourth digit. This difference turned out to be large enough to steer the simulation in a totally different direction.

In 1972 Lorenz introduced the term 'butterfly effect' to describe this phenomenon. In a lecture he then stated that 'a butterfly's wing flap in Brazil can cause a tornado in Texas'. The comparison was put in place for him by a colleague. Lorenz himself had used a seagull in his version of the analogy, in an unspecified place, and had confined himself to an ordinary storm.

The conclusion is clear. If such tiny differences have such a dramatic impact, long-term weather forecasting is doomed to fail. Even if the model were exact, the data from which we start will always contain measurement errors. These observations form the basis of a whole new branch of mathematics, chaos theory. In 2014 the Brazilian Artur Avila won the Fields Medal (the highest mathematical award) for his contributions to this field.

a great deal of knowledge of that specific application. We trust the predictions because we have full insight into the formulas and the physical phenomena they describe.

Mathematical models and teamwork Mathematical models are often very useful to enable teamwork around concrete problems. Take the blast furnace again to produce liquid steel. Keeping such a complex installation up and running requires a lot of scientific expertise, first and foremost of the chemical processes. A chemical engineer has that knowledge and is often able to convert it into a mathematical model. A mathematical engineer, on the other hand, has the knowledge to use this mathematical model to calculate predictions. These two specialties can be separated: a large number of questions about accurate predictions can be tackled purely mathematically, without worrying too much about the chemical meaning of the equations. The same goes for the modelling of the weather: it is not necessary (and often not even possible) for the weather specialist and the applied mathematician to be the same person. So, teamwork is necessary. The mathematical model is the tool that allows people with very different specializations to talk to each other. Applied mathematics is anything but a solitary activity.

Black-box models It is sometimes very difficult to write down the exact relationships between the relevant quantities in a mathematical model. This is, for example, the case for the electricity grid. We do have Kirchhoff's laws representing the continuity of electrical current, but the grid is so extensive and so complex that it is very difficult to simulate in every detail. Moreover, the voltages and currents on the network depend on the users and their behaviour is difficult to predict on the basis of simple physical laws. For example, no psycho-sociological law has so far been discovered that describes very precisely how quickly people turn on the light when it gets dark a little earlier due to bad weather. What we do have is a very large amount of data about the demand for electricity in the past. On the left of Fig. 13.1 you can see the electricity consumption in Belgium on 5 consecutive days in the week of 6 March 2002. There is a clear cycle of 24 h with a morning and evening peak and a valley at night. On the right-hand side of Fig. 13.1, consumption is also shown as a function of time, for the successive weeks of a whole year. We see clear differences between weekdays and weekend days. There is also a gradual evolution over the year.

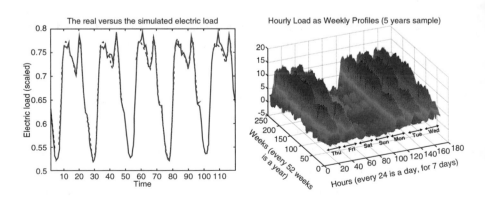

Fig. 13.1 Left: Time series of electricity consumption on 5 consecutive days in the week of 6 March 2002. Right: Electricity consumption for the consecutive weeks of a whole year. You see the hours of the week from left to right; the successive weeks from front to back. The height of the graph shows the electricity consumption

The challenge now is whether, without knowing the underlying laws, we can also create a mathematical model based on the available measurements. Can we – in a purely mathematical way – make predictions using only data from the past? We call such a model a *black box* model: we put in the black box (that we cannot open) a certain point in time and it returns an estimate of the power consumption at that moment, but how exactly this is calculated is hidden from us. For all we know, they might have appeared by magic inside the black box.

How are black-box models built? Of course, there is no magic inside the black box. Techniques from machine learning are also used to build models to make predictions. Just as with automatic fraud detection and automatic classification, as already explained in Chap. 12. However, there is one extra complication: in this case, the measurements are not independent of each other, but they immediately follow each other in time. The algorithms have to be adapted a little, but when it comes down to it, the techniques we described before – such as artificial neural networks – can be perfectly used in this context as well. After all, 'machine learning' is nothing more than the algorithmic (and mathematically based) selection of models that match the measurements as closely as possible. Here, too, training is needed. We initially feed the black-box model with measurements of power consumption over a certain period of the past and ask it to predict further (but still in the past). By comparing the prediction with the real measurement – which we also have – we can adjust the model so that it comes closer to that measurement. A bit like children learn after a while how long it takes for their plate to fall to the ground when they throw it away. And also here we have to keep some of the data aside to check whether the model is well trained. (Otherwise we can't trust the results of the black-box model).

The further ahead, the less accurate Electricity consumption fortunately shows no chaotic behaviour, as the weather does, but periodic behaviour on three levels: a day, a week and a year. This makes forecasting much easier. Yet even in this case, predictions become less accurate as we look further into the future. Figure 13.2 shows the final power consumption and the predicted use with a well-trained black-box model. On the left we see the first 5 predicted days. On the right we see an equally long period, but 2 weeks later. The difference in the quality of the forecast is clear.

How sure are we of predictions? The mathematics to produce predictions has more or less matured by now. In most cases, we can routinely produce predictions, both with models derived from physical principles and models derived

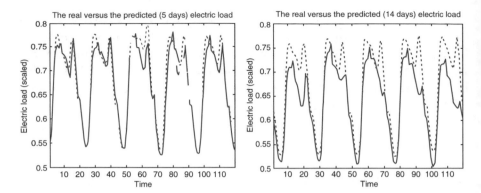

Fig. 13.2 Left: Time series of electricity consumption on 5 consecutive days in the week of 13 March 2002 (full line), together with the forecast based on the data in the week of 6 March (dashed line). Right: Time series of electricity consumption on 5 consecutive days in the week of 27 March 2002 (full line), together with the forecast based on the data in the week of 6 March (dashed line)

from machine learning. But we have also seen that the quality of the prediction decreases the further we look into the future, and also that the extent to which that happens depends very much on the specific situation. (An accurate weather forecast of more than a week is completely impracticable, while the predictions of electricity consumption still look reasonable even after 2 weeks). The main mathematical (and algorithmic) research therefore focuses on not only being able to make a prediction but also on being able to indicate how reliable that prediction is. When you watch the weather forecast, the uncertainty is displayed on screen, at least as far as the predicted temperature is concerned. It is easy to see that this is important information. A weather forecast is only valuable if it is correct with a probability of more than 70%. If it is not, it is more accurate to just assume that tomorrow's weather will be the same as today!

The impact of uncertainty on power generation For power generation, forecasting total demand is a little easier than a weather forecast. Unfortunately, there are also quite a few uncertainties on the production side, especially now that more and more renewable energy is being used. One of the difficulties arises from the massive emergence of solar panels. To estimate the electricity production of solar panels correctly, we first and foremost need an accurate weather forecast. In addition, we need to know exactly how many solar panels there are and where exactly they are located. After all, solar panels do not only influence global electricity production. They can also have important local effects, such as a local overload on a high-voltage line. It is therefore very important for the grid manager to have a good overview of this. This is also the reason for the high fines – up to 25,000 euros in Belgium! – for not declaring solar panels correctly.

Who will be the next Nostradamus?

There are many different techniques to make predictions based on existing time series. Some of them are based on artificial neural networks, but by no means all of them. Even within the class of artificial neural networks there are many choices. There are different ways to define the 'best' model. We have many options for choosing the precise nonlinear functions in the neurons. Not all methods calculate their predictions at the same speed. It is therefore not surprising that there is a lot of competition between researchers who each call their method the 'best'. A classic trick then is to make results public for just those examples that highlight the advantages of your method and make the other methods look bad.

In order to get some objectivity in this, Spyros Makridakis started a competition in 1982, which was later continued in 1993 and in 2000. The purpose of the competition was to test prediction methods for their generality, speed and accuracy. Therefore, the participants were provided with time series from very different situations: measurements of an experiment with a laser, physiological data of a patient with sleep apnea (such as heart rate, EEG, oxygen concentration in the blood), and stock prices. One of the tasks was even to complete a fugue that Johann Sebastian Bach had not yet completed when he died in 1750.

Meanwhile, there are dozens of competitions that challenge researchers to test their prediction algorithms.

The need for accurate predictions occurs in a wide variety of situations. GPS products are increasingly trying to predict the traffic jams along your route at the time you will pass by – rather than working with the traffic jams that are already there. To safely steer large ships in the harbour, it is necessary to predict water currents. Forecasting sea and river water levels is very important for flood defence systems, for regulating dams and avoiding flooding. A good prediction of the future value of a share on the stock exchange can also be very profitable. As soon as we have extensive time series from the past, forecasts can be made using black-box methods.

We have to make a few important remarks about methods like these. Firstly, they are not always suitable for long-term predictions. (That is the case in practice, and there is also theory to support that statement). Moreover, they are generally only useful when we have sufficient data about the situation we want to predict. The sheer fact that we predict the future from the past directly implies that we cannot predict events that never happened in the past. The Lebanese-American writer and philosopher Nassim Taleb describes this in his book 'The Black Swan' as follows: how do you describe the chance to meet a black swan if you have never seen one? His conclusion is that it is more important to make predictions robust in the face of unexpected events, rather than somewhat more precise in normal circumstances. Predictions are called robust if they still yield reasonable predictions in very extreme (and unlikely) situations. Therefore, when we consider robustness to be more important than precision, we attach more importance to reasonable predictions for a wide range of possible scenarios, rather than to a very precise prediction for only the most likely scenario. This position is understandable, especially when you know that Taleb was a financial speculator in a previous life. However, it does not always hold true: for example, in the case of the blast furnaces, unexpected events are well under control and the precision of the prediction is really crucial. So the 'best' method is not the same for every problem.

Part III

Mathematics for Tomorrow's Society

Chapter 14
Do Smart Girls Stay Single Longer?

When women start looking for a steady relationship, education turns out to weigh heavily. In Europe, a large majority of women choose a partner with a level of education at least equivalent to theirs. Now that a majority of university students are female, this poses some interesting questions. Are there still enough suitable men for all these female students? When will these women find the right one? And when will they start having children?

What's the problem? These kinds of questions are at the heart of *demography*, the science that deals with the evolution of populations. In short: birth, death and migration. There is quite a bit of historical information on this topic, such as censuses. These are supplemented by more qualitative data, often obtained through surveys such as the European Social Survey. These contain information about the intentions of individuals, such as their desire to have children, partner choice and migration (Fig. 14.1). To give meaning to these large amounts of data, we need mathematical models: formulas that suggest links between relevant quantities (such as age, duration of a relationship, etc.).

Many small ones make one great Demographic evolutions are a result of a large number of individual decisions. After all, each young woman decides for herself when she has found a suitable partner and if (and when) she wants to start having children. The rules for these choices differ from person to person and are a combination of many (sometimes unconscious) factors. For demography, all those individual choice processes are not that important in themselves. The aim of modelling demographic processes is to gain insight into

G. Samaey, J. P. L. Vandewalle, *The Invisible Power of Mathematics*, Copernicus Books, https://doi.org/10.1007/978-1-0716-2776-1_14

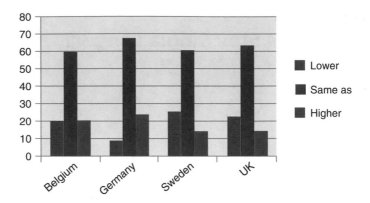

Fig. 14.1 Percentage of marriages in which the man is lower, equivalent or higher educated than the woman, for four European countries

evolutions of the whole population, not into the partner choice of individual women. The only important thing then is to be able to make statements about the percentages of women who choose certain types of men and the age at which they do so. We make abstraction of all characteristics that have no influence when looking at the entire population. Physical characteristics, for example, or character traits. The aim is to consider only those factors whose influence we want to study. In this case, age and educational level of potential partners.

Molecules and dating How can we correctly make abstraction of irrelevant properties? This is a problem that has been studied mathematically in many other scientific domains, such as chemistry and physics. For chemical reactors, for example, much is already known about this. Microscopically speaking, a reactor vessel that produces a certain chemical product is filled with a large number of individual molecules that react with each other. Each of these molecules has a number of properties that are important, such as its species and location. (In general, there are only a few types of molecules, and only a limited number of reactions are possible when two molecules meet.) Other properties, such as the speed or orientation of the molecule sometimes play a less important role. We often can make abstraction of these properties of individual molecules. The behaviour of the entire reactor can then be described by a relatively simple macroscopic model, in which only the *concentrations* of the substances in the reactor are important. The chemistry between people is – fortunately – much more complicated. Contemporary mathematicians are therefore still busy investigating how existing derivations – such as those for simple chemical reactors – can be generalised to more complex systems such as those arising in demography.

From chemistry between molecules to chemistry between people Now the question remains how the behaviour of individuals should be described. It is impossible to represent all social interactions in real life in full mathematical detail. Fortunately, that is not necessary either. We can come a long way by simply casting some important patterns into rules and applying those rules on a collection of fictitious persons in a computer simulation.

For the example of partner choice, we can start from the following (strongly simplified) assumptions: women want a partner with more or less the same level of education, and of a similar age. For men, it (unfortunately) turns out to be more realistic to assume that their partner's level of education does not play a strong role and that the ideal age of the partner is 25 years, regardless of their own age.

A formula for attraction

Let's look at the attraction that a woman with age c and education level F feels for a man with age d and education level G. We are looking for a formula for the attraction A that is maximal when the man has the same age as herself and also the same education level. In all other cases it should be lower. Because the woman's age is fixed at c, we can write the attraction as a function of the man's age. We then look for a function like in Fig. 14.2.

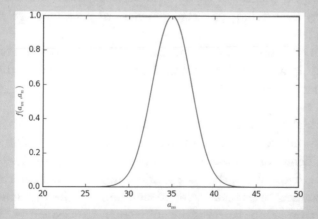

Fig. 14.2 Function representing the attraction of men to a 35-year-old woman, as a function of the age of the men. In the model, men of the same age are the most attractive. The attraction decreases strongly (and symmetrically) with the age difference, and approaches zero when the age difference becomes too large

(continued)

In that figure, the age of the woman is taken to be $c = 35$. So the attraction is maximal if the age of the male is $d = 35$. As soon as d deviates, the attraction force is lower, and there is almost no attraction anymore when the age difference is too large. The attraction must not become negative either. A function that meets all these requirements is, for example,

$$f(c,d) = \exp\left(-\lambda \cdot (d - c)^2\right)$$

This formula contains a parameter λ that determines how quickly the age difference makes a man less attractive to the woman in question. Evidently, a similar formula and reasoning can be given for the difference in education level. There is also an Achilles heel for this kind of models. It is very difficult to determine model parameters, such as the parameter λ above, accurately. Moreover, such parameters may be different for each woman. How can we know how much an age difference affects the attractiveness of a potential partner? To obtain estimates of these parameters, statistical methods are applied on the available demographic data. The result is then not the 'correct' value of the parameters, but their 'best estimate' based on the available data.

Onto the dating scene As soon as we know how to determine the attraction between two people, we can draw up a model for partner selection. The rules of the game are as follows: we create a social network with virtual men and women that indicates which people know each other;

- Depending on the size of their social network, each person has a certain probability of meeting potential partners of a certain age and level of education;
- In such an encounter, there is a probability that 'a spark will occur' and that a relationship will arise. The greater the perceived attraction, the greater the chance that the spark will occur.

Exactly which encounters take place – and whether they result in a relationship – is determined by chance in the model. In the simulation, this is controlled by generating random numbers according to the probability distributions involved. So, there are random numbers that determine at what time an encounter takes place, with whom, and whether the spark occurs or not. Each simulation is therefore different. Consequently, information obtained in this way cannot possibly relate to the specific lives of the individuals. However, we can obtain statistical information about emergent behaviour, such as the average age at which people enter a relationship, how long it lasts, etc.

How reliable are those mathematical models? Now that we know that demographic processes can be simulated on the computer, one important question remains: what can we actually conclude from these simulations? To answer that question, we need to reflect for a moment on the two types of assumptions we made when drawing up the model. First of all, the model assumes that age and level of education are the characteristics that determine attractiveness. But at the same time, we also make an assumption about *how strongly* these effects are taken into account. The model contains parameters

---

Online dating according to mathematicians: OkCupid.com

Individual partner choices require more detailed models than the prediction of demographic evolution. OkCupid is an online dating site that goes to great lengths in determining the potential attraction between two individuals.

OkCupid was founded by a number of mathematicians and asks each participant to complete an extensive questionnaire. You not only give an answer for yourself but also the answer you expect from a possible partner. The latter is very important. Think, for example, of a question like 'Do you like to be the centre of attention?'. Someone who answers 'yes' to that question will probably be better off with a partner who prefers to stand in the shadows than with a partner who is also looking for attention.

A big problem with OkCupid is converting all questions to a score (in percent) that indicates how well two people would fit together. Two ideas are used here. First, each partner chooses a *weight* for each of the questions. Next, OkCupid calculates the weighted average of the answers in a specific way.

There are many ways to calculate an average, of which the 'ordinary' average (which we call 'arithmetic average') is just one. We then take as the mean the sum of N numbers and divide that sum by N:

$$\frac{X_1 + X_2 + \ldots + X_N}{N}$$

OkCupid chooses the geometric average: the average of a row of N numbers $X_1, X_2, \ldots, X_N$ is then given as the N-th root of the product of those numbers:

$$\left(X_1 \cdot X_2 \cdots X_{N-1} \cdot X_N \right)^{1/N}$$

You can easily check that this formula leads to a lower value than the arithmetic mean if the numbers X are *very different* from each other. This is because the geometric mean is never greater than the arithmetic mean and is only equal when all numbers are identical. To see this, you can, for example, compare two situations with two questions: in one case, you have a match of 50% on both questions, and in the other case, you have a match of 0% and a match of 100%. For the arithmetic mean this makes no difference, but the geometric mean will

(continued)

be 0% in the second case. OkCupid uses the geometric mean exactly for that reason: when choosing a partner, it is much better to have a reasonable 'match' on all levels than a perfect match on half of the questions.

that determine how quickly a difference in education level or age results in a reduced attractiveness. These parameters strongly influence the results of the simulations, but unfortunately they cannot be measured. Mathematical models can therefore not in themselves serve as an explanation for a certain phenomenon. The results must always be viewed together with other data.

Simulations test possible explanations Simulations are currently mainly used to determine whether scenarios are plausible, and whether or not a certain factor is relevant. For example, a clear trend can be observed in the European Social Survey data: nowadays, fewer women marry a man with a higher level of education than their own, compared to a few decades ago. There are (at least) two possible explanations for this. One explanation fits perfectly in our romantic image of our time: the higher level of education of women – and the associated financial independence – ensures that women are freer in their choice of partner, and therefore assign a lower importance to the future income of their husband. The other explanation is more prosaic: women less often marry men with a higher level of education because they are simply less available.

We can test both explanations via simulations. We carry out two simulations: one in which only changes are made in the percentages of women and men per education level, and a second one in which the preferences of the women are also adjusted. The simulations with unchanged preferences appear to be much closer to the measurements. When preferences are also adjusted in the model, the simulations predict an even stronger decrease in the number of women with a higher educated partner than is actually the case. Therefore, there is no reason to assume that women are less picky now than they used to be on the diploma of their future partner.

Moreover, there are other data that indirectly support this conclusion. Indeed, it is striking that in larger countries (such as Spain or Germany) remarkably fewer women marry a man with a lower level of education than their own, compared to small countries such as Belgium. It is probably no coincidence that students in these large countries – because of the distance – spend the whole semester at the university campus, while in Belgium students return home every weekend. This is a hint that the frequency of encounters has a larger influence than an individual's preferences: people tend to fall in love with people they meet. It would be interesting to figure out how dating apps influence the above story.

And are smart women going to stay single longer? That women choose a steady partner at an increasingly later age is clear from the available measurements. The age at which women become mothers for the first time increases even more drastically. However, it is very difficult to determine whether this is a consequence of the changed gender relations within the highly educated population segment. It is very well possible that women still meet their dream partner as early as before, but that they are not yet open to a serious relationship at that moment, for example, due to study pressure. Or that they postpone their first child because of their professional career. Both the woman's preference and the availability of suitable partners play no role in these scenarios. In order to verify this, more complex mathematical models will be needed that take more factors into account. The research for this is now in full swing.

How many school-age children can we expect in the coming decades? How many retirees will there be and how big is the active population? What impact will migration have on our population structure? The answers to these (and other) questions have a major societal impact. They determine to a large extent what public investment is needed and how our social security system can be financed. Accurate forecasts are therefore crucial in order to make the right choices today. A task in which mathematics has a major role to play.

Chapter 15
What If There's More Data Than Storage?

571 new websites per minute. 48 h of new videos on YouTube. 684,478 Facebook posts. Over 100,000 tweets. Those were the numbers in 2012. According to cautious predictions, the annual production of digital data doubles every 2 years. That data needs to be stored. The availability of storage space grows much more slowly. Inevitably, storage will run out if we are not careful. But how can we store all this data most efficiently?

What's the problem? We live in a world where digital photos, videos, sound recordings, biomedical data, biometrics and many other data can easily be measured, sent and stored digitally. Estimates for the year 2010 indicate that the entire world produced some 1250 billion gigabytes (or 1.25 zettabytes!) of data that year. These are more bits than there are stars in the universe. Every year more data is added, and even the rate at which data is generated keeps increasing. It turns out that every year there is about 58% more data stored worldwide than the year before. Our internet traffic is not the only cause of this. A lot of data is also generated by sensors. For example, technology producers are strongly committed to the continuous measurement of all kinds of biometric data in order to encourage and promote a healthy lifestyle. Think, for example, of the new fad of 'smartwatches', wristwatches that continuously measure your heart rate, count your steps and much more. This trend is called the *Quantified Self* and provides an enormous amount of data that needs to be stored and processed. More and more sensors are also being built into your home and car, for example, for energy efficiency or temperature control.

On the other hand, the total amount of storage capacity – on hard disks, DVDs or other media – is 'only' increasing at a rate of 40% per year, partly simply by manufacturing more storage space, but also by developing new technology. Of course, an annual increase of 40% is not nothing, but because

G. Samaey, J. P. L. Vandewalle, *The Invisible Power of Mathematics*, Copernicus Books, https://doi.org/10.1007/978-1-0716-2776-1_15

data production is increasing faster, we can still get into trouble. A first milestone was reached in 2007, when as much new data was produced in a single year as there was storage capacity available worldwide. In 2011, twice as much data was produced as there was storage capacity. The discrepancy between the production of data and its storage is increasing and processing capacity cannot keep pace with data production either. Nevertheless, no major disasters occurred in 2007. How is this possible? There are two main reasons. Firstly, of course, we do not have to keep track of all the data simply because they were generated. (Although, this happens increasingly often because we have more and more technology at our disposal to use the data in a useful way). But – and this is much more important – we can also retrieve a whole arsenal of mathematical tricks to represent the data more efficiently so that they take up less space than seems necessary at first sight.

Magic with zeros

To be able to talk conveniently about gigantic amounts of data, we briefly go over the most important jargon. The basic unit of digital information is a *bit* (a binary information unit that can take the value 0 or 1). When we combine 8 of those bits, we obtain one *byte*. A byte can thus represent the numbers 0–255 (because $2^8 = 256$). With a kilobyte, you probably think of 1000 bytes, but that's not quite right – it's $2^{10} = 1024$, an extra 2.4%. (A computer prefers to calculate with powers of 2, and it just so happens that the tenth power of 2 is about equal to 1000). A megabyte is then 1024 kilobytes, and so on. We give an overview of the most common prefixes, with their classical and binary meanings.

Symbol	Prefix	Standard meaning	Binary meaning	Difference
k	kilo	$10^3 = 1000^1$	$2^{10} = 1024^1$	2.40%
M	mega	$10^6 = 1000^2$	$2^{20} = 1024^2$	4.86%
G	giga	$10^9 = 1000^3$	$2^{30} = 1024^3$	7.37%
T	tera	$10^{12} = 1000^4$	$2^{40} = 1024^4$	9.95%
P	peta	$10^{15} = 1000^5$	$2^{50} = 1024^5$	12.59%
E	exa	$10^{18} = 1000^6$	$2^{60} = 1024^6$	15.29%
Z	zetta	$10^{21} = 1000^7$	$2^{70} = 1024^7$	18.06%
Y	yotta	$10^{24} = 1000^8$	$2^{80} = 1024^8$	20.89%

By the way, the largest number with such a name is the 'googol', namely, 10^{100} a 1 followed by 100 zeros. The founders of Google were so fascinated by this number that they named their company after it.

The difference between data and information Let's start with a common method of representing text, especially the ASCII code. The ASCII code is useful to work with but not very efficient in terms of memory usage. The ASCII code represents each uppercase, lowercase letter, digit and punctuation mark by a number between 0 and 128. Because there are still some digits left, we still have some margin to add some 'control symbols', such as the 'beginning' and 'end' of a line or the whole text. Seven bits are needed to represent those 128 symbols. In the ASCII code, one byte (8 bits) is chosen. So, an eighth bit is added. This eighth bit is used to detect possible errors in the binary representation. A text in ASCII code is thus represented as a very long sequence of bits, grouped by 8 in bytes, with each byte representing one character.

But the fundamental question is whether all those bits are useful. It's not that hard to see that some letters are more common in English than others. The 'e' and the 't', for example, are very common, while the 'q' is hardly ever seen. (How often a letter occurs can also be deduced from the numerical value that Scrabble assigns to it). Moreover, letters cannot occur in random order because we always have to form existing words. For example, when a word starts with 'th', we know that, with a high probability, a vowel or an 'r' will follow. And if you look at the next sentence, it is not difficult to fill in the missing letters:

I h.v. a dr...m th.t one d.y th.s n.ti.n w..l r.se up.nd live o.t the t.ue m..n..g of its c..ed

By simply taking into account English glossaries, sentence structure and the meaning of the sentence, you notice that you hardly have to gamble to recognize the famous words of Martin Luther King Jr. The missing characters are actually determined completely by the available characters and context. So they might as well be left out, because they don't contain any new information. That we write them anyway, is because we like to make language redundant. We make sure that there is data on surplus to make texts faster and easier to read. This helps, for example, when someone writes so illegibly that we cannot decipher some characters.

How much information does a letter contain? This immediately raises a natural question: how many bits on average do you really need to encode a single character? How much information does one character contain on average? To find that out, the American Claude Shannon devised an experiment. He had a number of participants complete an English text, letter by letter. The first character was given. From the second, the participants had to guess the next character. When only one guess was needed, Shannon concluded that that letter contained no additional information. It was perfectly predictable. According to Shannon, when two guesses were needed, the letter contained one bit of information (as there appear to be two possibilities). And so on. After repeating it with a large number of test participants and texts, this experiment shows that an average of 2.4 gambles are needed – a surprisingly low number. Because $2.4 = 2^{1.25}$ Shannon concluded that it should be possible to present a text in the English language with an average of 1.25 bits per letter. Those 2.4 guesses are an average. In the beginning of a sentence generally more gambling is needed. At the end of long words often less.

Find and remove errors?

Something can always go wrong when storing or sending information digitally. A hard disk can contain errors here and there, where certain bits can no longer be read or are simply wrong. A DVD can be scratched. And when sending data over the internet, there can be a malfunction on the line. Each of these problems causes bits to be read incorrectly. Especially with efficient compression, this can make decoding seriously difficult or (in the worst case) completely impossible. That's why we always provide an extra piece of information that is not really needed to represent the content, but that only indicates that the message has been sent correctly. Many countries, for instance, have a check algorithm for bank account numbers, to ensure that money is transferred to the intended account. In Belgium, this check algorithm works as follows: a bank account number consists of 12 digits, and the last two digits are obtained by dividing the first ten digits by 97 and taking the remainder. It can be proved that any single-digit typo will result in an invalid account number, as will most of the two-digit swaps.

With the ASCII standard, the error detection is done by adding an extra bit. The ASCII code consists of 128 symbols, which requires 7 bits. For example '1100101' for the 'e'. However, each of these symbols is represented by one byte (8 bits), where the eighth bit is used for error detection. The idea is simple: the eighth bit is chosen in such a way that each byte contains an even number of zeros and ones. For the letter 'e', the binary code then becomes '11001010'. We call the eighth bit a parity bit. Now, when one bit is transmitted incorrectly – and so a 1 becomes a 0, or vice versa – the number of ones in the byte will be odd. We then know that an error has occurred. Only we don't know which bit is wrong! To also *correct* errors, we can invent more complicated codes. An example is to use every eighth byte into an additional *parity byte*. To avoid using up too much space in this book, we illustrate the idea here with blocks of length 4 instead of bytes. We send three blocks (each with their own parity bit, here displayed underneath each other). Then we send a fourth block that ensures that the number of ones is now even in each column, as shown below:

0	1	0	1
1	1	1	1
0	1	1	0
1	1	0	0

(continued)

Three characters are transmitted here using four blocks of four bits. Each character is 3 bits long, and the fourth parity bit ensures that the sum per row is even. The fourth 'character' is not part of the message, but ensures that also the sum per column is even.

What happens if an error occurs in one of the bits? Suppose, for example, that the second bit of the first character is wrong. The parity bit of that first letter will then indicate that that character is wrong. The parity bit at the bottom will then tell you that it is the second bit that is wrong. It is now a piece of cake to correct the error.

0	0	0	1
1	1	1	1
0	1	1	0
1	1	0	0

Of course, there are much more powerful error correction and error detection methods that can also detect errors in highly compressed files. For example, the above principle can be generalised to a 'checksum': we then calculate a number (the checksum) from the original message before compression. During decoding, the same checksum is then recalculated to verify the message.

Compression without loss: zipping documents The next challenge now is to present a text in such a way (other than with ASCII) that, in the end, only about 1.25 as many bits are needed as there are letters in the text. Just because Shannon demonstrated that such a thing is possible in theory, doesn't mean we immediately have a code that actually achieves this efficiency. A first idea to compress text is conceptually very simple: don't represent all letters with the same number of bits, but represent letters that occur more often with a small number of bits. Letters that occur less often are then represented with more bits. This idea originates from the Morse code from 1835, which is popular among radio amateurs. The Morse code is also a digital code, since all letters are represented as a sequence of two symbols, namely 'short' ('.') and 'long' ('-'). Those short and long signals can be sound or electric current, or something else. Since the letter 'E' is the most common in English, it is given a short code, namely '.'. The other short code ('-') represents the other most common letter 'T'. The infrequent letter 'Q' has a long Morse code, namely '--.-'.

The best-known code of this type is the Huffman code. Huffman designed it during his PhD at MIT in 1952. The anecdote is that Huffman's supervisor, Robert Fano, left his students the choice between an exam or homework. The task looked fairly simple: find the most efficient way to create a variable-length digital code. Huffman found it a rather tempting prospect not to have to study for the exam and set his sights on the homework. He didn't know at the time that Fano and Shannon themselves had been searching in vain for the most efficient code for several years. When Huffman finally gave up and started studying, he suddenly got his idea, which was so brilliant in its simplicity that Fano exclaimed desperately: "Is that all it took?" The idea? First look at which letters are the least common, give them their (long) code and work your way back to the shorter codes, instead of the other way around.

Today, the Huffman code still forms the basis for the popular 'zip' compression, together with another idea from 1977: the Lempel-Ziv code. Here a text is simply kept from the beginning and we look at it letter by letter to see if we've encountered it before. Take the following sentence: '*It was the best of times, it was the worst of times.*', by *Charles Dickens*. We can now start storing this sentence from the beginning. As soon as we notice that the phrase 'It was the' has already been stored, we don't need to store it again. Instead, we refer to the last time we came across this phrase. The two numbers we need for this are the place where the phrase 'it was the' first started (the first character) and its length (11 characters, spaces included). We then replace the phrase '*it was the*' with those two digits. We can do the same with the words "of times" that also appear twice. So we write "*It was the best of times, (1,11) worst (16,8).*". We now continue like this: from now on, we replace every sequence of letters that we already encountered by only two digits, a starting position and a length. In this way, quite a few fragments of text can be represented by just two digits, independent of the entire length of the fragment. Clearly, with much less literary appeal.

Compression with loss: MP3 With music, we have more freedom than with text because here it is not really necessary to reconstruct the original sound signal perfectly. It is enough to ensure that the signal we recreate *sounds* the same. All inaudible information in the sound signal can be thrown away perfectly, nobody will notice. But what exactly do people hear? Clearly, we don't notice all the information in sound. For starters, people don't hear frequencies higher than 20,000 Hertz. We have already stated for the cochlear implants, on page 35, that it is sufficient to store slightly more than 40,000 numbers per second of sound: two per frequency. For music, we choose 44,100 numbers per second. Each number is represented by 2 bytes (16 bits), which gives us 705,600 bits per second. We have to do that twice, for stereo. So, we end up with 1,411,200 bits per second. A song of 3 min then would require a solid 67 megabyte. That's still way too much.

For music (and other sound), the MP3 standard is the most commonly used compression technique. MP3 manages to store a song of 3 min in a file of about 6 megabytes, a reduction by a factor of 11. The technique was developed in 1993 in the German Fraunhofer Institute, under the direction of Karlheinz Brandenburg. The compression algorithm uses a well-known psycho-acoustic phenomenon: sound masking. This phenomenon describes how a perfectly audible sound – such as a television set that is switched on – can become inaudible when a strong other sound source is created – such as someone who starts vacuuming. Music is full of signals that become inaudible due to the simultaneous presence of other signals. Compression methods can exploit this by filtering out frequencies that are inaudible from the signal and throwing them away during compression. This is called 'perceptual coding'.

Compressing photos and films We can also apply all these principles to photos and films. Just like text, images contain a lot of repetition. For example, many photos are taken outside under a bright blue sky. The top rows of the image often contain the colour 'blue'. Just like text, we can now encode with two numbers: one for the colour blue, and one to indicate that the next 100 pixels have that colour. With film, we can do even more: normally, consecutive images in a video are very similar, with only differences where something is moving. Even then, only the changing pixels need to be saved. With this kind of compression techniques, it is of course very important that everyone compresses in exactly the same way because without proper agreements on compression, the reconstruction of the original image or movie (decompression) becomes completely impossible. That is why people work hard on standardization. For photos we use the JPEG standard, for video MPEG. The development of such a standard is done by a consortium in which all major makers of recording and playback equipment are present. This requires years of work and is subject to both commercial and strictly technical interests. An important milestone was reached in 2001 with the H.264/MPEG-4 AVC standard, the standard used for blu-ray, among other things. It also describes in great detail a number of technicalities. One example is motion compensation, in which motion can be predicted from successive frames. In that case, the whole motion no longer needs to be stored, only the motion that deviates from the predicted one.

Compression of data is extremely important, not only for texts and entertainment but also for many social and economic applications. In the medical sector, for example, data is generated constantly. Electrocardiograms, medical images and brain scans often show largely normal patterns, with only minor deviations. Because only this anomaly contains useful information about possible diagnoses, this type of information can also be highly compressed for more efficient storage. Hospitals therefore invest heavily in specific data storage systems (often under the name 'Picture Archiving and Communication System' or PACS), in which an entire infrastructure is also provided for security and communication.

The abundance of security cameras and sensors of all kinds also generate an avalanche of data that is barely controllable. People are therefore working hard on methods to perform the compression during the measurement itself. The aim is to curb the flow of data at the source. The underlying reasoning is that most measured signals and images are not random, but from a limited group of 'possible images'. As soon as we know this collection of 'possible images', all we have to do is point out the (already stored) image, which requires very little memory. This way of working does raise some conceptual questions. Can we trust the scientific conclusions and the evidential value of images when there is no trace of the actual measurement data? Can we tell when a decoding algorithm is infected with a computer virus? And can we still trust measurement data and images in a court of law? These are all questions that will require a lot of research.

Stelarc, the Man with the Third Arm

The Australian Stelarc, born in Limassol (Cyprus) in 1946, is a performance artist who explores and shifts the boundaries between man and technology. In his performances he overcomes the limitations of his body by means of prosthetics – not to replace badly functioning body parts, but to *expand* the possibilities of his body. With him, the prosthesis no longer symbolises a deficiency, but rather a surplus.

In 1980, Stelarc finishes a work that will strongly determine his career: *Third Hand* is a robotic arm, attached to his own right arm and operated by electrical signals generated by his leg and abdominal muscles. With his three hands at the same time, he writes the word *Evolution,* thus clearly illustrating the blurring boundaries between man and machine. Stelarc performed with his *Third Hand* until 1998. Since then he has gone even further. For *Fractal Flesh* he developed a system with electrical muscle stimulation to control his body remotely over the internet. He also grew an extra ear with a built-in microphone on his left arm – also connected to the internet.

Such works of art cannot be created without a healthy dose of mathematics. *Third Hand* needs special software that converts muscle contractions into an electrical signal that controls motion. The motion itself also needs to be described mathematically so that the robot hand can perform it perfectly. This requires a lot of signal processing and geometry. The same technology is extremely important for serious applications, such as active prosthetics – prosthetics that replace amputated limbs and can be used as a real arm. An important example is a functional robot hand for accident victims. The technology is also used for remote operations, where a surgeon controls a robot hand over the internet. (This is useful, for example, for delicate operations for which not many surgeons have the necessary skills).

The questions posed by Stelarc's work are universal: What do technological adaptations to our bodies mean to our humanity? How does technology determine our identity? Can our bodies be better adapted to the high-tech environment in which we live? Without mathematics we would have none of this: neither the high-tech environment nor the art to think about it.

Chapter 16
Driving Without a Driver

Cars are doing more and more by themselves. Whether it's comfort, safety or communication, more and more tasks in the car are taken over by electronics and computers. Already quite a few cars can park independently, keep a distance from a car in front of them or adjust when the driver deviates too far from his driving position. Soon, the Tesla S will even be able to automatically overtake slower cars – and that with a simple software update! Google has been testing completely driverless cars for several years now. It is expected that half of the new cars will be self-driving by 2032. But how do such driverless cars work? And how can they revolutionize our mobility?

What's the problem? Many of the problems associated with car traffic are directly related to human limitations. For example, more than 90% of accidents involving cars are caused by human error. A moment of distraction, calling (or even texting!) behind the wheel, too tired, drunk, in a hurry, etc. The list of causes is endless. Even without accidents, the human driving style is not optimal. We drive alone in cars with room for a whole family, causing traffic jams. And we often drive too closely to the cars in front of us, which means we have to brake too fast and too suddenly in an unexpected situation. That too can cause traffic jams, and it's not good for fuel consumption. Can self-driving cars eliminate accidents? And what is their impact on fuel consumption and congestion problems?

Following the road independently A first step on the way to a fully self-driving car is to let the car follow the road without colliding with surrounding cars. An experienced human driver does this by estimating their own position and speed and considering the position and speed of the cars around them

G. Samaey, J. P. L. Vandewalle, *The Invisible Power of Mathematics*, Copernicus Books, https://doi.org/10.1007/978-1-0716-2776-1_16

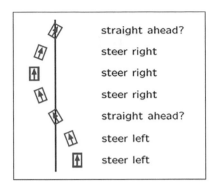

Fig. 16.1 Schematic representation of the control mechanism for a car trying to drive straight ahead

(and also the course of the road itself). Based on this, the driver – for a large part even unconsciously – makes a prediction of the situation within a few seconds if they don't intervene. For example, they will notice that they won't take a turn correctly if they don't steer, or that they will hit the car in front of them if they don't hit the brake. The driver then uses that estimate to take action: steer or brake (Fig. 16.1).

When a computer has to drive the car, it has to have the same information at its disposal. The computer can know on which road it is with a map and a GPS, and it can be equipped with sensors to measure its precise position on the road and its speed. The surrounding vehicles, for example, can be detected with cameras. But then comes the difficulty: a computer cannot automatically estimate what the situation will be within a few seconds. This has to be described with mathematical models; in this case, the basic laws of physics. They indicate unambiguously how a turn of the steering wheel is transformed into a change of direction of the car, and how the speed of the car changes with a pressure on the accelerator pedal or the brake. When we then set a trajectory and speed, the computer can calculate how to control the steering wheel and accelerator to travel this trajectory.

How the car should react exactly depends on the speed: at a lower speed, you turn differently than at a higher speed. But also the type of car plays a role. In 1997, Mercedes released the first model in the A-Class. That model turned out to tip over during the so-called moose test, a test to see if the car can perform an evasive manoeuvre without any problems – like when in Scandinavian countries a moose suddenly appears in the middle of the road. So, a specific mathematical model will be needed for each type of car.

Driving in the city When a car drives around in the city, many more factors play a role than just steering. The car has to stop at traffic lights, let pedestrians cross at pedestrian crossings and overtake cyclists safely. It becomes even more difficult to take other road users into account if they themselves do not follow the rules: a self-driving car, for example, must also be able to stop quickly when a child (or an adult) suddenly crosses the street at a place that is not intended for that purpose. But things must also remain a bit practical. It cannot be the intention to stop next to every pedestrian, just to make sure that they do not accidentally walk under the self-driving car.

An experienced driver considers the speed and body language of the pedestrian, as well as certain environmental factors, when estimating this type of situation. In this way they make an implicit prediction of what will happen in the next few seconds. Unfortunately, these kinds of difficult traffic situations do not allow themselves to be expressed so quickly into a few simple rules. How can a driverless car mimic that behaviour? How can a car correctly interpret all incoming signals?

Just like mathematical prediction on page 101, it boils down to letting the driverless car 'learn' how to interpret a certain situation. For this purpose, an artificial neural network will once again be able to prove its usefulness. As soon as the computer is able to make predictions, its control algorithm can be used again. Moreover, there is one important observation to make: it is (fortunately) not necessary to 'learn' to drive every car again – an important difference with human drivers! As soon as a self-driving car has been trained enough, the model to predict the behaviour of the environment can be used in *all* self-driving cars! Currently, Google's self-driving cars together have already covered more than two million test kilometres, and they share all their experience with each other. Each self-driving car contains more than 40 years of human driving experience, so to speak!

How safe are self-driving cars already? For safety's sake, the test drives with Google's self-driving cars are still done by a human test pilot who can take over in case of problems. The pilot also notes situations in which the car has difficulties deciding. The test drives show that self-driving cars have problems with exactly the same situations as human drivers, e.g. a pedestrian trying to cross but deciding to let the car through first. The moment the car has braked as well, one of the two has to let the other one through, and that sometimes takes some doing. Another difficult case is deciding whether to swerve or brake in the event of a sudden obstacle.

In the early days, Google only did test drives on closed circuits, and the intention was to avoid cones and dolls. In that kind of testing, it's relatively harmless to really explore the limits of what's possible, and so the self-driving car was programmed in an aggressive driving style. This led to running engines, squeaking brakes and terrified trainees in the passenger seat. The software was designed in such a way that the level of aggression of the driving style could be adjusted. On public roads, the cars are set to drive very carefully, perhaps best compared to someone taking driving lessons. It is therefore not surprising that there are more problems with annoyed human drivers than with the self-driving cars themselves. Because of this caution, the number of real accidents with self-driving cars is very low. In the period from 2009 to 2015, only 16 accidents were reported: three in the early days when the software hadn't really learned enough, and 13 after that. In those last 13 accidents, each time the self-driving car was hit by a human driver who was a little too enthusiastic with their phone and forgot to pay attention to the car in front – something you cannot blame the software for.

From cruise control to driverless car

We illustrate the principles behind the adjustment of a system first for a simple example: the cruise control that is ingrained in many cars. The intention is that the car will drive at a fixed speed without the driver having to press the accelerator. This not only leads to less tired feet on long journeys but also to a significant reduction in fuel consumption. By the way, the first car with (mechanical) cruise control was the Chrysler Imperial from 1958.

At first sight, this problem seems too much simplified to derive a general method for self-driving cars. However, we will see that the same principles are important for steering, turning and keeping distance from a car in front. Even the estimation of the traffic situation does not appear to be fundamentally different as soon as we formulate the problem in a slightly more abstract way.

But first the concrete problem of cruise control. Here, our goal is to regulate the fuel supply to the engine in such a way that the car always drives at the same speed. If we drive uphill, the car will release more fuel. If we drive downhill, the car will use less fuel. To achieve this, the car measures the current speed, and adjusts the fuel supply based on the measured difference between the actual speed and the desired speed. For the sake of convenience, let's assume that we have to accelerate. The slower the car drives, the more it will have to accelerate. But how much exactly? If we accelerate too much, we will go too fast and we will have to brake immediately after. If we don't accelerate enough, it will take forever to reach the desired speed. So, there is clearly an optimum. But even if we don't execute that optimal way of accelerating perfectly, there is normally no problem. The cruise control is stable and will recover in normal circumstances. Although you might imagine that the system could derail if you first accelerate far too hard to brake immediately and eventually have to accelerate even harder. So the mechanism of cruise control has to be properly adjusted somewhere.

In order to use the same principles for the other aspects of self-driving cars, we first draw the working method a little more abstract using a block diagram, see Fig. 16.2:

(continued)

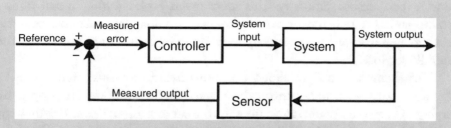

Fig. 16.2 Schematic representation of a block diagram to think about automatic control

We consider what we can measure as the 'output' of a 'system'. The 'system' here is the car, and the 'output' is the speed. The desired speed is then the 'desired' output of the system, i.e. the reference. The 'measured' output is the current speed. If a system has an output, it also has an input. We call the 'input' in our example the flow rate of the fuel supply. We now add a 'controller', whose task is to keep the speed constant. For the fuel supply, this is the accelerator pedal. The regulator adjusts the 'measured output' of the system by adjusting the input of the system so that the measured output changes in the direction of the desired output. In this case: how to accelerate.

The great strength of this block diagram lies in its generality. The only thing we have to change to apply this control technique to other aspects of the self-driving car is our definition of 'input' and 'output'. Instead of a speed, the output then becomes the desired trajectory, and at the input, we have the position of the steering wheel. The rest of the method remains identical! This block diagram is generally used in almost all applications where a certain desired state must be maintained, such as cooling in chemical reactors and nuclear power plants or trajectory control for rockets and satellites. Even the toilet sink is refilled in this way!

Impact on heavy traffic In addition to safety, congestion is also an important factor in our car use. Here, too, it is precisely the imperfect human driving style that causes traffic jams. This has been the subject of very controlled experiments. In 2008, the Japanese researcher Sugiyama and his colleagues published the results of such an experimental study. They let 22 cars drive on a circular road 230 metres long, and asked the drivers to drive neatly one after the other, at a fixed distance and at a fixed speed. Of course, the drivers can't perform this task perfectly, and so they adjust their speed: a little acceleration or a little braking. The experiment clearly illustrates how accordion traffic jams occur spontaneously in heavy traffic. One small anomaly (a driver suddenly having to brake his car for the vehicle in front of him) causes the car behind him to brake as well. The car behind him also has to brake a little harder, so the car behind him also has to brake a little harder. The fifth car in line is then almost stationary. This mechanism explains why, in real traffic, traffic jams occur behind drivers who perform inappropriate manoeuvres, such as aggressively and incorrectly overtaking or continuing to drive on the middle section.

A mathematical analysis shows that self-driving cars cause traffic jams in the same way when they control their speed in the same way as people: by trying to keep the distance constant with the car in front of them. Fortunately, self-driving cars can also do other things that are much more difficult for people. For example, trying to stay halfway between the car in front and the one behind them. People cannot do this because they can only look in one direction at a time, but a self-driving car can use all its sensors at the same time. In this way, self-driving cars could immediately become a lot less sensitive to traffic jams.

Won't we just spend more time in our cars? Self-driving cars will make our lives a whole lot easier. They'll be safer than cars with human drivers. We'll be able to make more use of our time in the car. They'll also be less likely to cause traffic jams. And they are also more economical than an ordinary car: they drive more defensively, keep the prescribed distance and brake less abruptly in the event of unexpected events. So they avoid exactly those aspects of human driving that drive fuel consumption into the air. So many advantages that we have to worry: won't we all use the (driverless) car so often that we nullify all its advantages? Won't there be more self-driving cars than there are regular cars today, and won't they all travel more miles?

It takes more than a little math to solve these problems. Creative business models will be needed, for example, by making car sharing more convenient. Existing systems of car sharing are only a limited success, mainly because we all have slightly different needs. For example, we might have to return home in the evening from work with a different stopover, which will get us into trouble if we drive with a colleague in the morning. With a self-driving car, car sharing becomes much easier: you can probably imagine an app that indicates a time when you want to leave for work the next day. At that time, a self-driving car will arrive, which will already carry a number of other passengers, each with their own destination.

That car sharing can provide greater efficiency. Currently, driving accounts for 18% of the budget of an average family, while we only spend 5% of our time doing it. Calculations by (again) Google show that a system with self-driving taxis can work at an occupancy rate of 75% without users experiencing many delays. And a study on car traffic in Lisbon indicated that a tenth of the current number of cars could suffice without having a negative impact on people's mobility. (This is because the time people have to wait for a self-driving taxi is more than compensated by the fact that they spend less time in traffic jams). Collaboration with trains – with large parks of self-driving cars that pick up train passengers at home – is also an option.

New design requirements for self-driving cars A driverless car consists not only of software but also of hardware – the car itself. It is, of course, perfectly feasible to make an existing car self-driving. (The current Tesla S, for example, is even equipped to drive autonomously via a simple update of the software). But from the moment you know that a car is meant to be self-driving, you can already take that into account in the design phase. That way, the car no longer needs to have steering wheels or pedals, nor is it really necessary to put the seats in the driving direction. In addition to these practical elements, mathematically more difficult design requirements are also influenced. For example, the shape and choice of materials of a self-propelled driverless car no longer have to be determined solely by its crash resistance. Self-driving cars can be much safer than cars with human drivers, but they are still fully responsible for the accidents they cause. Of course, crash resistance is still important: nobody wants to be in a car that dents like a packet of butter. But self-driving cars can take more account of accident *prevention* in their design, such as, for example, an optimal shape to allow the sensors to record as much information about the environment as possible.

Driverless cars will force entire industries to transform. Car builders will have to adapt: instead of creating products for buyers, they will have to focus on offering car use as a 'service' for sharers or subscribers. The insurance market will also have to adapt. There will be far fewer accidents, but with the accidents that do occur, it will be unclear who is responsible. The driver? The other party? The creator of the software? There are also a number of ethical dilemmas associated with this question. Although there will be fewer deaths on our roads with driverless cars, those deaths will be caused by an automatic choice by a computer program. Who can we punish then? Will the damages be recovered from the car manufacturer, and thus included in the purchase price? And there are even more difficult questions. The driverless car makes a choice between the type of accident. Will it choose to run over the child on the zebra crossing, or make an evasive manoeuvre in which the driver dies? Can you include the 'economic value' of a possible victim in that decision? What if there is still a small chance that you brake in time (and therefore do not have to swerve)? And who buys a car that sacrifices the life of its owner for that of a pedestrian? All these questions will require a thorough social debate.

Chapter 17
How Vulnerable Are Our Banking Systems?

The financial crisis that started in 2008 brought down several banks and led to a prolonged global recession. Many blamed the mathematical models used to determine the value of financial products. But is that right? Was the crisis caused by mathematics? And can mathematics be used to see crises coming or even prevent them?

What's the problem? The financial system is the engine of our economy. It allows, for example, to postpone payments through loans, to hedge risks through insurance or to invest in promising companies through shares. In all these activities, it is important to set the right price for the financial product and thus to be able to properly assess future profits and risks. But it doesn't stop there. Risks do not disappear through such transactions. They are only transferred from one party to another. That other party will in turn also want to protect itself and has a whole arsenal of appropriate financial constructions at its disposal, which we call financial derivatives. In studying the opportunities and risks involved, we automatically enter the domain of financial mathematics.

Exchanging risk with forward contracts Let's start with a simple example. Say you are a European company that exports to America. We'll take the Belgian brewing company InBev as an example. U.S. buyers pay InBev in dollars on delivery, but InBev needs euros. Then InBev can agree a *future* (or futures contract) to convert the dollars it receives from the customer into euros at the current exchange rate *at the time of payment*. In this way, InBev avoids making a loss if the dollar drops in value in the meantime. There is, of course, a cost associated with such a contract: if the dollar had actually gained value, InBev would have made an extra profit, and it is now losing out on the possibility of that extra profit.

Of course, the risk of the dollar dropping in value has not disappeared. It has only been transferred from InBev to the counterparty. The counterparty

© The Author(s), under exclusive license to Springer Science+Business Media, LLC, part of Springer Nature 2022
G. Samaey, J. P. L. Vandewalle, *The Invisible Power of Mathematics*, Copernicus Books, https://doi.org/10.1007/978-1-0716-2776-1_17

must deliver euros to InBev at the agreed price, which may exceed the value of the dollars at that moment in time. The counterparty will therefore generally be someone who wants to protect itself against the opposite risk, i.e. an increase in the value of the dollar. This could be, for example, a company that imports into Europe from America. Such a company would want to avoid the opposite risk. We take as an example a distributor of Apple products. If that Apple distributor and InBev enter into a future contract, they both have certainty about the value of the foreign currency they will receive in the future. They therefore have no risk. If the dollar were to decline in value, the Apple distributor would sacrifice its additional profit to offset InBev's loss. Conversely, in the event of an increase in the dollar, InBev sacrifices its additional profit to offset the loss of the Apple distributor. In this way, financial instruments create a win–win scenario: both parties sacrifice an opportunity for additional profit to avoid a risk of loss.

Options are a bit more complicated We can go a step further, for example, by looking at *options*. When you buy an option, you don't pay for a product itself, but for a certain *right to* that product, for example, the right to buy or sell that product for a certain price in the future. To return to our example: the exporting Belgian company can also opt to buy an *option* to – at an agreed time in the future – exchange dollars for euros at a predetermined rate. In this case, you don't use the option when the dollar rises in value and simply collect your extra profit: more euros for the same amount of dollars. The cost for InBev then consists of the purchase price of the option. The seller of the option has to solve two problems: they have to put an appropriate price on the option and they have to define a strategy to cover their own risk of a strong depreciation of the dollar.

The value of an option and the Black–Scholes model The problems with options were not solved until 1973, when American mathematicians Fischer Black and Myron Scholes published a mathematical model for the value of an option based on the expected future value of the underlying asset. Shortly thereafter, Robert C. Merton provided the mathematical interpretation that allowed the Black–Scholes model to become widely used. Because the future value of an asset is subject to uncertainty – precisely the risk against which we want to hedge – it cannot be determined precisely and certainly, but only in a statistical sense. That is why we speak of the *expected* future value. The Black–Scholes model allows the seller to define a strategy to eliminate their own risk. Roughly speaking, the seller of the option must constantly monitor the value of the underlying asset and buy and sell this asset at appropriate times.

Difficult to obtain, easy to use In addition to the fact that it is relatively reliable, the success of the Black–Scholes formula can also be explained by its ease of use. Although the model is based on deep mathematical concepts that quantify uncertainty, the end result was a fairly simple formula. So simple that a few months later it was added by Texas Instruments to their pocket calculators. The Black–Scholes formula is rightly regarded as the cornerstone of today's financial mathematics. Scholes and Merton were awarded the Nobel Prize in Economics in 1997. Unfortunately, Fischer Black had already died by then.

The limitations of the Black–Scholes model Of course, this model doesn't solve all financial problems: as often, the devil is in the details. In this case, those 'details' are the assumptions made in deriving the model. It starts with the parameters: the Black–Scholes equation uses the *future volatility of* the underlying asset. Without becoming too technical, we can consider volatility as the extent to which the value of the underlying asset fluctuates in a directionless manner due to accidental market fluctuations. However, the problem lies in the word 'future'. Of course, we cannot know the future volatility. Moreover, the Black–Scholes model assumes this volatility to be constant over time, just like other parameters such as the interest rate. It does not take into account the cost of transactions, nor the closing of the stock exchanges. All these simplifications lead to unforeseen risks. It should therefore come as no surprise that – to this day – mathematicians are in the process of improving, expanding and refining the Black–Scholes model (Fig. 17.1).

Trade between computers and 'flash crashes' Mathematical models such as the Black–Scholes model can be a good guide for professional trading on the stock exchange. However, it is important to know the limitations of the mathematical models to avoid surprises. At the same time, a large part of the trade on the stock exchanges is carried out entirely by computers, without human intervention. This phenomenon is called *algorithmic trading* and is used to automate large volumes of relatively small transactions, hoping to control and spread risks. Often, automatic traders only keep products in possession for a few minutes or seconds, sometimes just a few milliseconds. On some exchanges, some 40–70% of transactions already take place without human intervention. In general, this goes extremely well, but if things go wrong for a while, the consequences can be enormous. On 6 May 2010, for example, the stock exchanges in the so-called 'Flash Crash' fell very quickly (up to 9% in a few minutes) (Fig. 17.1). In those few minutes, some 1000 billion dollars were lost. How exactly this happened is unclear – and is still the subject of a lot of ongoing research. One of the hypotheses is that the system had become

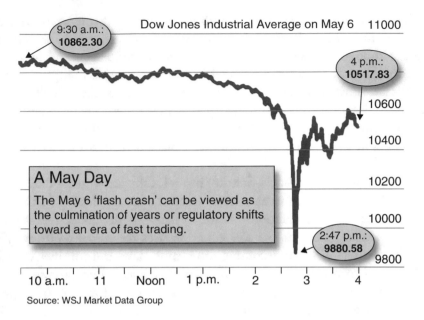

Fig. 17.1 Flash crash of the stock market on May 6, 2010

unstable for a short period of time due to a downward spiral in the reactions of algorithmic traders to each other's transactions.

Individual risk and systemic risk Although the Flash Crash cost some people a lot of money, its damage was repaired within a short period of time – and some people probably made a good profit. Real disasters only happen when careless risk management puts the entire financial system at risk. At the end of 2008, this happened due to a complex interplay of factors, including the false assumption that the value of homes in the United States would continue to rise. As a result, among other things, the value of financial derivatives based on mortgages on these homes was overestimated. However, an unexpected risk on those financial products does not automatically lead to a global financial crisis. What circumstances, and what successive events, then convert an individual risk into a risk for the entire financial system, a *systemic risk*?

Reducing individual risk can increase systemic risk The last word has not yet been said about that question, not in the least because it is very difficult to give a precise, quantitative definition of systemic risk. Much exploratory research is currently being done into mechanisms that cause systemic risk, and analogies with better understood physical systems are frequently put forward. An example of such a simple analogy was proposed in the research group of the Greek-American mathematician George Papanicolaou (Stanford), based on a phase transition in a magnetic material. In this qualitative model, individual banks are represented by particles moving in a mountainous landscape (the red ball in Fig. 17.2). There are two valleys, the left valley where things are going well economically, and the right valley where things are going badly. Each individual bank moves back and forth in its valley under the influence of market fluctuations. If such a fluctuation is large enough, there is a chance that the bank will go over the top of the mountain into the other valley and go bankrupt. That is the *individual risk*. In order to avoid this risk, the banks link their destinies: their risk management strategies ensure that their financial situation evolves towards the average of the group of banks (the blue ball in Fig. 17.2). The *systemic risk* would then be that the red ball moves to the other valley, meaning that all banks would fail together, collectively. Such a transition is called a *turning point*, or a *catastrophic shift*.

Of course, such a model is too simplified to be directly applicable to reality, but its simplicity allows one to study certain mechanisms with mathematical precision and in great detail. For example, Papanicolaou was able to argue that, in certain cases, the strategy of each bank to attach itself to the others could increase the risk of all the banks moving *together to the other* valley. The strategy of banks to avoid their individual risk then increases the systemic risk, without them noticing it!

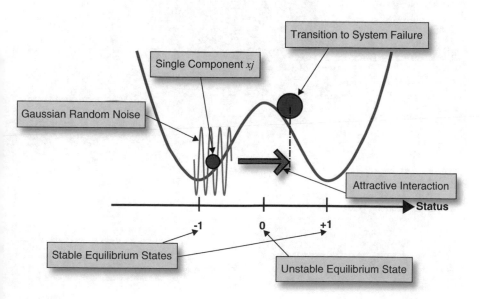

Fig. 17.2 Schematic representation of a system with multiple equilibria. In the left valley everything goes well. In the right valley there is a crisis. A system can suddenly end up over the hill in the other valley due to an accidental disturbance, if it is powerful enough

Tipping points

To understand a tipping point, we make a simple analogy: a marble rolling in a hilly landscape, for example, stretched by an elastic band. The marble represents the system (in our case, the banking system). The place where the marble is located is its condition (here the economic cycle). The elastic band is the environment: factors that influence the economy but are outside the banking system, such as employment or the political situation. In the middle of the five situations in Fig. 17.3, the elastic is stretched in such a way that the marble can remain in two valleys. For the banking system, these two valleys are the 'good' and the 'bad' economic state. Which of the two valleys the marble ends up in depends mainly on its starting position and speed. A transition between the 'good' and 'bad' economic state is only possible if a sudden external shock disrupts the position of the marble to such an extent that it reaches the top of the hill and can roll to the other side.

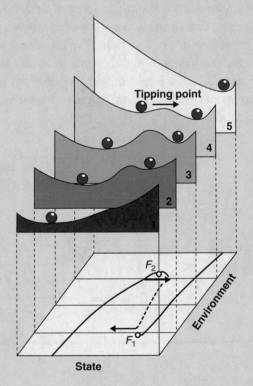

Fig. 17.3 A system with multiple equilibriums under changing environmental factors. A crisis may become more or less probable when the environment changes

(continued)

On the front and back figure, we change the shape of the elastic (and thus the environment) in such a way that only one valley remains. This is a metaphor for an economic and political climate that only allows for a boom or an economic crisis. The intermediate figures show the transition: in a changing political or economic situation, one of the valleys gradually becomes less deep, until it finally disappears. A marble lying in the disappearing valley then suddenly rolls a long way to the other valley, and this only required a very small change in the shape of the elastic. Worse still: when the elastic band regains its previous shape, the marble doesn't return to the valley it originally came from.

The figure at the bottom shows the possible positions of the marble (the economic state) plotted against the shape of the elastic (the environment). Full lines represent *stable equilibria*: valleys from which the marble does not roll away. The dashed line is an *unstable equilibrium*: a top from which the marble rolls away at the slightest perturbation. The figure shows how the position of the marble (the state) changes slightly when the shape of the elastic band (the environment) changes slightly. When the tipping point occurs, however, this leads to a large sudden change that cannot easily be undone.

We don't always have to wait until the environment has changed so much that a valley disappears completely. Characteristic of such dramatic changes is that they are often caused by an accidental disturbance. The figure also helps to understand this: in situations 2 and 4, there are still two valleys, but one of those valleys is not very deep. Even in these circumstances, a relatively small external shock can cause the marble to roll to the other side.

Ecology for bankers Papanicolaou's model is too simplistic in many respects. It makes no distinction between small and large banks, and it assumes that all banks are linked in an identical way with all other banks. In reality, the banking system consists of a mixture of many small and a few very large banks, and the whole banking system forms a complex network with dynamic interactions. The American ecologist Simon Levin (Princeton) made an analogy with another network, the ecosystem. In the case of the ecosystem, the network represents the species and how they depend on each other. The systemic risk is then the probability that the entire ecosystem collapses when a species disappears. Our ecosystem turns out to be fairly robust for disappearing species. Levin's research is therefore trying to find out which characteristics of the network give it its robustness and how the structure of financial networks differs from ecosystems. Here too there is still a lot of work to be done, but a paper in Nature from early 2008 (i.e. before the financial crisis hit) alluded to the possible systemic risk that could result from the American mortgage crisis.

Mathematical modelling is crucial for our economic system. When used judiciously, it allows individuals and companies to control their individual risks. However, mathematical modelling is also important for regulators, who are responsible for minimising the risk of collapse of the entire financial system. Research into the causes of major systemic risks in mathematical models therefore focuses on two main aspects: mechanisms that cause an inherent destabilisation of the financial system, but also mathematical techniques for predicting a catastrophic shift from data on the current evolution of prices before it actually takes place. Moreover, if the latter research is considered at a sufficiently abstract level, its results can also serve to predict major changes in ecosystems or in the climate.

Chapter 18
Can We Predict Climate Change?

The earth is heating up, and man is responsible for that. By the year 2100, the temperature on earth will be one to twelve (!) degrees higher than it is now, with all its consequences. But how do we know that? And why is there so much uncertainty on that prediction?

What's the problem? Earth temperatures in the year 2100 are predictions. They haven't been determined experimentally. (After all, we can't build a planet and run a whole range of lab tests on it.) Instead of experiments, we use simulations with mathematical models. These consist of formulas that show the relations between the relevant quantities such as temperatures of land, water, air, and ice, but also atmospheric pressure, atmospheric composition, and so on. But how do these simulations work exactly? And how reliable are they?

Going to Antarctica To understand climate change, we don't just need some mathematics. To arrive at a sound mathematical description of all the important factors, we need to start from measurements that show how the climate has worked in the past. For these measurements, climate researchers rely on the Antarctic ice, which is very pure because no people live there. In Antarctica, one can drill ice cores and see how the composition of the ice changes with depth. Because the ice thickens every year due to extra freezing water, the depth is directly related to time. (The deeper, the longer ago the ice formed). This makes it possible to accurately determine the composition of the water and the atmosphere in the past. The temperature at that time can also be deduced from the ice. To find connections between all these quantities, statistical methods are used, such as a calculation of the correlation between different quantities. For example, it has been shown that higher average temperatures and higher concentrations of CO_2 generally go hand in hand.

© The Author(s), under exclusive license to Springer Science+Business Media, LLC, part of Springer Nature 2022
G. Samaey, J. P. L. Vandewalle, *The Invisible Power of Mathematics*, Copernicus Books, https://doi.org/10.1007/978-1-0716-2776-1_18

Measuring and modelling hand in hand Taking measurements in Antarctica can give a lot of insight into the evolution of the climate so far (and can also be very adventurous). However, it is difficult to make predictions based only on these measurements. A much more systematic approach is to use our knowledge of the meteorological, physical and chemical processes in the atmosphere to create a dynamic model for the evolution of the relevant quantities over time. Such a dynamic model is in fact nothing more than a formula that calculates the value of those relevant quantities at some moment in time, based on the values of those quantities at an earlier point in time. For example, the temperature and atmospheric pressure tomorrow from today's temperature and atmospheric pressure. In this way, we then 'step forward' from day to day to make a prognosis in the longer term.

A simple model for the heat balance of the earth

The simplest climate model imaginable (and therefore not a particularly accurate model) considers only the average temperature on earth, where by 'average', we mean the spatial average over the whole globe, as well as the temporal average over the year. This model describes an energy balance between the incoming heat from the sun and the heat loss through radiation. We note the temperature at time t as $T(t)$ and obtain the following equations:

$$C \cdot T(t + \Delta t) = C \cdot T(t) + \Delta t \cdot \left(Q_{in}(t) - Q_{out}(t) \right)$$

where the constant C stands for the heat capacity of the earth, $Q_{in}(t)$ for the incoming heat from the sun per unit time, $Q_{out}(t)$ for the heat leaving the earth by radiation per unit time and Δt for the size of the step we take forward in time (e.g. a year). This equation expresses conservation of energy, but we cannot continue without further specifying a number of quantities. The incoming heat from the sun is given by

$$Q_{in}(t) = (1 - \alpha) Q_{sun}(t)$$

in which $Q_{sun}(t)$ represents the heat that the sun emits as radiation and α is a constant that indicates which fraction of that heat is reflected by the atmosphere. We describe the outgoing heat by the universal law of radiation. In the nineteenth century, the Slovenian Josef Stefan (based on experiments) and the German Ludwig Boltzmann (on purely theoretical grounds) concluded that the heat emitted by a spherical body is proportional to its temperature to the fourth power. That law can be written in formula form as

(continued)

$$Q_{zon} = (1-\epsilon) \cdot \sigma \cdot T(t)^4$$

where σ represents a universal physical constant (the Stefan–Boltzmann constant) and ϵ a parameter indicating the fraction of heat retained in the atmosphere. The constants α and ϵ are determined by the composition of the atmosphere and are, just like C and $Q_{sun}(t)$, estimated from the available measurements.

If we now write the temperature change per unit of time as the difference between the incoming and the radiated heat, the trained eye may see the introduction of the derivative as a measure of the instantaneous temperature change. This will lead to the introduction of differential equations as a modelling technique. Although this model is very simple, it allows human influences to be taken into account. Greenhouse gases ensure that less heat can leave the earth. In the model, this effect can be represented by increasing the parameter as a function of time. This relation is then determined by how the amount of CO_2 increases, something that will require an additional model of the same kind.

On land, at sea and in the air Climate models must be able to provide more information than just the evolution of the average temperature on earth. Indeed, temperature rises are not the same everywhere and are accompanied by many other phenomena, such as the melting of ice masses and the corresponding rise in sea level, but also changes in precipitation patterns, such as more frequent and larger storms or long periods of drought. To accurately describe such phenomena, the models must include a large number of relevant quantities (such as temperature of air, land and water, but also atmospheric pressure, humidity, etc.). In addition, because of their dependence on the location on earth, the models must calculate these quantities not only as averages but also *locally*, at any point in the atmosphere and oceans, and at any point on the surface of the earth, as a function of time.

In practice, we limit ourselves to the average value on a large number of rectangles on the earth's surface for a large number of discrete points in time. Even with this limitation, we soon find ourselves with many millions of calculations. Only a computer can perform such calculations, but the computation is so heavy and extensive that the resolution of the current simulations is limited to rectangles of 10 by 50 km even on the most powerful supercomputers. And even then, one simulation takes on average a few months, on a supercomputer that can perform 10^{16} operations per second by connecting hundreds of thousands of processors.

The computational complexity of climate simulations

The fastest current supercomputers use several million processors and perform about 30 billion operations per second (that's a 3 followed by 16 zeros). And yet those computers calculate for months to predict the climate of the next hundred years on a grid of 10 by 50 km, with about 100 layers in the atmosphere. How much faster does a supercomputer have to get to perform the same simulation on a 1 by 1 km grid, with 200 layers in the atmosphere?

All in all, the refinement does not seem too drastic, but there are immediately 1000 times more grid points for which we want to calculate the relevant quantities (10 × 50 × 2). Unfortunately, the amount of calculation does not increase proportionally to the number of grid points. The calculations are done in steps: in each step, the next state at a point in time some time Δt later is calculated based on the current state. Let's assume that the amount of calculation per time step increases like the square of the number of grid points. This is the minimum when you have to perform a calculation for each grid point that uses information from all the other grid points. This estimate is generally very optimistic, but in this case the calculation time increases by a factor of four when the number of grid points doubles. When we then have 1000 times more grid points, the amount of calculation increases by a factor of 1 million. In addition, refining the grid by a factor of 10 means that the maximum time step Δt becomes at least 10 times smaller. (A cloud that moves, for example, will get into a neighbouring rectangle 10 times faster if the rectangles are 10 times smaller). So the computer must take 10 times more time steps to bridge the same time span. In total, a rather limited increase in resolution therefore immediately requires 10 million times more computing power.

Clearly, waiting for faster computers won't save us. Computational mathematicians are therefore working intensively on new simulation algorithms. They look at different avenues: reducing the amount of computation per unknown variable, reducing the number of unknowns and increasing the maximum time step. Over the last 50 years, the total gain from faster algorithms has been at least as big as the gain from faster hardware, and better algorithms are still being developed. Which will undoubtedly continue...

The devil is in the details   The resolution with which climate simulations are carried out today is not fine enough to represent details such as individual clouds. Mountain ranges can also be a problem. However, such details have a major impact on the results obtained! Just think of nearby valleys where it is raining in one valley and bright weather in the other. The rain clouds can 'hang' behind a mountain chain. To capture such effects, climate specialists add a number of parameters to the model. These parameters are then set in such a way that a number of selected observations are reproduced correctly. Because experts have different opinions about the importance of the various observations, no less than 24 different models are used in reports by the Intergovernmental Panel on Climate Change IPCC.

The only way to stop the discussion is to perform calculations with a much higher resolution. But that's exactly where the problem lies: to make calculations with rectangles with a side of 1 km possible, you need a computer that can calculate millions of times faster than the fastest existing supercomputers. Fortunately, there is a branch of mathematics, computational mathematics, in which new algorithms are systematically developed that can accelerate such calculations up to thousands of times without significant loss of accuracy.

Uncertainty reigns! In addition to the lack of resolution in the simulations, the presence of tipping points also plays an important role. We already met tipping points on page 141, when we discussed the stability of the financial system. We also encounter tipping points when studying the climate, for example, in the case of global ocean currents. These currents are driven by temperature differences and differences in the salinity of the water. They redistribute heat over the earth's surface and thus have a stabilising effect on the climate. The largest current of this type is called the 'large ocean conveyor belt' because of the large amounts of heat it transports. It consists of a warm surface flow (red in Fig. 18.1) and a cold deep water flow (blue).

The large conveyor belt is not fixed. Depending on the temperature and salinity of the seawater, the strength of the flow may shift. In general, these changes are small and occur very gradually. However, a small change in temperature can cause an unexpectedly sudden change, i.e. a tipping point. One possible mechanism is the following. As the temperature rises, the ice on Greenland melts faster and faster. This is how a large quantity of cold water ends up in the North Atlantic Ocean. All models indicate that such a large quantity of cold water can cause the direction of the large conveyor belt to reverse. This will happen quite suddenly: from a certain critical temperature, the ice melts sufficiently fast to realise this reversal. When the large conveyor belt reverses, this will have a dramatic effect on the climate. At the moment, Belgium and Western Europe, for example, have a much milder climate than North America due to the supply of warm seawater. If the large conveyor belt is reversed, this situation will be turned around. Small shifts can therefore lead to major consequences.

Unfortunately, most mathematical models for the climate have been drawn up based on data for the current direction of the large conveyor belt. With our

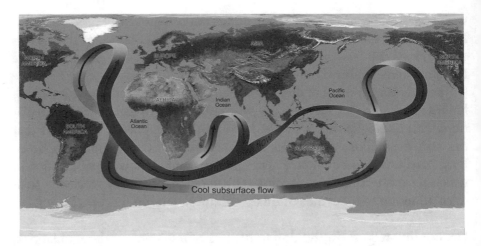

Fig. 18.1 The large ocean conveyor redistributes heat over the earth's surface. A sudden reversal of these currents can have drastic consequences for the climate

knowledge of physics, we can predict the mechanism leading to a reversal of the large conveyor belt, but it is very difficult to predict exactly the critical temperature at which this will happen. Together with the rather imprecise description of very local effects, this explains to a large extent the great uncertainty on long-term climate predictions.

Climate change will affect all aspects of our daily lives: low-lying land will disappear below sea level, large areas will become uninhabitable or infertile due to drought or storms, resulting in large migratory flows. Measures are needed to control climate change – and the problems it causes. However, the large uncertainty in current forecasts makes it difficult to correctly assess the impact of concrete measures. Sceptics use this uncertainty (wrongly!) to question climate change itself! Better predictions of climate evolutions are therefore of invaluable social importance to convince the world of the seriousness of the situation and to correctly assess the impact of certain measures. Every day, new mathematical methods for faster computer simulation are being researched all over the world.

On Rubber Ducks and Other Toys

In millions of bathrooms, bright yellow bath ducks are lying around as decoration or as children's toys. But did you know that, for over 20 years, such bath ducks have been collecting precious information about one of the most important phenomena affecting the climate, the ocean currents?

On 10 January 1992, a container ship in the middle of the Pacific Ocean got caught in a storm on its way from Hong Kong to Tacoma (Washington). Due to the waves and the wind, a container got pushed off the deck and ended up in the water. The cargo, including 29,000 yellow bath ducks, ended up in the ocean. A few months later, the first ducks washed ashore in Alaska, but in the years that followed they ended up everywhere, from Australia to South America and the United Kingdom.

Oceanographers saw a great opportunity here. They have long had mathematical models of ocean currents and use them to research climate change, as well as the distribution of marine waste. However, their simulations are subject to considerable uncertainty because they have to start from a fully known picture of ocean currents at a certain point in time. Unfortunately, they only had a limited number of measuring points and had to take a guess for the missing data.

The rubber ducks gave an unexpected new twist to the research: if you can use simulation to predict where a rubber duck will wash ashore that has gone overboard in a specific place, then you reverse the computation! Then you can calculate back from where the bath ducks washed ashore and determine their most likely initial state.

This technique, known as solving *inverse problems*, is used more generally: if there is some missing information in a mathematical model, you can try to use available measurements to calculate the most likely values for your missing information. You can do even more: as data keeps coming in, you can adjust your estimates on the fly.

So it should come as no surprise that in the meantime much more is being counted than just rubber ducks: information has been used from, among others, 61,000 washed ashore Nike sports shoes (which by the way were still wearable after drying!), 100,000 toy cars, 34,000 hockey gloves and 5 million Lego bricks.

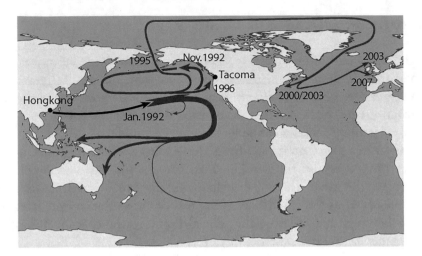

Chapter 19
War and Peace

Can we understand what mechanisms lead to arms races? Can governments calculate whether it is wise to start a war? And can we use those calculations to make the world a more peaceful and safer place?

What's the problem? Both in wartime and in peacetime, new weapons are being developed with the aim to have an advantage when it comes to an armed conflict. The use of mathematics for this purpose is obvious. More striking is that this context also gives rise to new fundamental questions about strategic decisions that also have their importance in a normal economic context. Moreover, mathematics is also used to evaluate the different options in strategic decisions. How many weapons are needed? Is it interesting to start a war? Can we safely agree with our rivals to disarm together?

Mathematics and technology in WWII The use of every conceivable technology to win a war has always been the nature of humankind. Technology (and the mathematics behind it) certainly was a decisive factor in the Second World War. The unlimited investments (in both camps) in technology such as radar, sonar and atomic bombs determined the outcome of the war, as did the development of the computer to decipher encrypted messages. Moreover, these inventions drastically changed life in the peacetime after 1945. Today's society relies heavily on technology developed during World War II: computers, nuclear power, jet engines for airplanes, and so on.

The development of this technology was usually based on existing mathematical insights. To give an example: the Russian mathematician A.A. Markov (1856–1922) had theoretically studied a class of stochastic processes and demonstrated how they could be used for the mathematical description of language. Later, this proved to be useful within the Manhattan project for the description of nuclear chain reactions in an atomic bomb.

G. Samaey, J. P. L. Vandewalle, *The Invisible Power of Mathematics*, Copernicus Books, https://doi.org/10.1007/978-1-0716-2776-1_19

Theoretical impact of war mathematics In addition, the technological revolutions during World War II also gave rise to a whole set of theoretical questions that gave direction to fundamental mathematical research from the 1950s onwards. Completely new domains were explored: computer science, information theory, Monte Carlo simulation, operational research and statistical quality control are only the most striking examples.

Take operational research, with the central question of how available material can be optimally used for a given purpose, taking into account logistical and budgetary constraints. Answers to these questions are of great importance for the planning of warfare, as well as for the logistical planning of many companies. During wartimes, it is especially important to come up with a reasonable solution quickly (and not to continue searching for the very best solution until after the war is over). After WWII – freed from the pressure to answer immediately – more time could be taken for the fundamental theoretical underpinning. When is a problem well formulated? What is meant by an *optimal* solution? And how can it be calculated? These questions have occupied applied mathematicians for decades, and still today we obtain new insights into this problem.

Can war itself be calculated? That mathematics is used in the development of military technology is clear. But can mathematics also be used to make strategic decisions during a war? In decisions about armament strategies in peacetime? Or in deciding whether or not to start a war? Opinions differ: the recent book *Mathematics and War* contains two successive chapters entitled *"War cannot be calculated"* and *"Warfare can be calculated"*.

There are many factors that make war unpredictable. The cohesion, fighting spirit and courage of soldiers in a war cannot be measured in peacetime. The strength of the opponent is not exactly known. Small but inevitable accidents or unforeseen circumstances can have a major effect. And sometimes it is not even clear who actually won a battle.

Nevertheless, attempts to model wars mathematically are made, often using probabilistic calculations for the outcome of a fight – a little bit like throwing dice in Risk. The chance of victory then depends on many parameters, such as the armament and strategy of both parties, the weather or the terrain. Such models are often used in peacetime for simulations of possible conflicts. The aim is then to look at different options for the strategic purchase of weapons systems. An example is the Danish *Defence Dynamics* model, which was used for long-term planning by both the Danish government and NATO from the 1980s onwards.

Figured security and arms races The armaments strategy of a country depends to a large extent on the (perceived) strategy of the others. In the 1930s, British mathematician and pacifist Lewis Fry Richardson (1881–1953) was concerned that an arms race between the European superpowers would lead to another great war. He developed a simple model, the parameters of which he determined on the basis of data on arms expenditure in the run-up to WWI. The model contained an equilibrium in which arms levels remain fixed, but this equilibrium is unstable: the slightest disturbance irrevocably gives rise to an arms race that cannot be stopped. Of course, the purpose of this kind of models is not to make very precise predictions. We should be happy if we are able to use them to make a number of qualitative statements about the nature of the events that can lead to war.

Controlled disarmament treaties Based on Richardson's work, a number of extensions have been proposed. The model has been extended to take account of the uncertainty that countries have about each other's level of armament. A distinction can be made between different types of weapons. And, of course, the effect of a war can be simulated by having each country destroy opponent's weapons. Even the risks of instability due to mutual disarmament in the SALT (Strategic Arms Limitation Treaty) agreements of the 1970s have been quantified with such models. The simplest way to take this last aspect into account is to consider three scenarios: 'our country attacks first', 'our country attacks second', and 'no one attacks'. Of course, we want the latter scenario to be optimal. (Fortunately, no one has gotten it into their head to implement the conclusions obtained with the very first models, which were way too simple. These models suggested to attack quickly, and to offer immediate help after an 80% destruction of the opponent. Of course, the model assumed that the opponent would still be able to act rationally then, and would use the same model).

The weather, the arms race and math – the world of Richardson

The fact that Richardson was a convinced pacifist can be seen, among other things, in his work on weather forecasts. In 1922, he published a ground-breaking work in which he proposed a method for weather forecasting based on a mathematical model and pure arithmetic. He wanted to release the work only after he reassured himself that 64,000 'computers' would be needed – at that time, computers were people sitting at a table calculating – to predict the weather in time for the next day. So, he thought that his method could never be of practical use in wartime. A small judgmental error when looking at the computing power of today's (electronic) computers!

(continued)

Let's now turn to Richardson's model for arms races. That model considers the armaments levels of two countries. Country 1 has an armament level $x(t)$ that evolves as a function of time t; for country 2, the armament level is denoted by $y(t)$. The armament level evolves on the basis of three effects. Firstly, each of the two countries has certain (historical) grievances against the other, on the basis of which arms are produced. Secondly, each country produces an amount of extra weapons that is proportional to the opponent's current armament level. Finally, the growth is slowed down by the fact that part of the budget also has to be used for the maintenance of the existing weapons capacity. The cost of this maintenance is proportional to the current armament level. All this leads to the following system of ordinary differential equations:

$$\begin{cases} x' = g + a \cdot y - b \cdot x \\ y' = h + c \cdot x - d \cdot y \end{cases}$$

with g and h the grievances, the coefficients a and c the degree to which both countries look at each other, and b and d the maintenance costs of their own armaments. Unfortunately, the coefficients in this model cannot be known. For example, the armament efforts of the opponent cannot be known exactly. Nevertheless, the models do allow one to derive conditions for military stability. Armament levels are balanced when they are equal to

$$x^* = \frac{a \cdot h + d \cdot g}{b \cdot d - a \cdot c}, \qquad y^* = \frac{c \cdot g + b \cdot h}{b \cdot d - a \cdot c}.$$

Such a balance is clearly a desirable situation, but having such a balance is not enough. We actually want something extra: if the balance is disturbed, we want the system to return to it. If that does not happen, an unbridled arms race will occur. For Richardson's model, this instability arises when countries react strongly to each other and have few economic constraints, a condition that holds when $ac > bd$.

Chaos, uncertainty and instability in international security An important military principle guaranteeing the stability of a peace situation is a balance between attack and defence expenditure. Defence expenditure ensures that an attack by the other party can cause less damage, whereas attack expenditure is aimed precisely at harming the other party. Because an effective defence against nuclear weapons is technically impossible (and research into it excessively expensive), the nuclear powers during the Cold War could only use attack weapons, and a potentially very dangerous arms race arose inevitably, in which the two blocks saw no other option than to systematically arm themselves more strongly in response to the arming of the other.

At one point, the intention was to stop this race by building complex defence systems such as the *Strategic Defense Initiative*, the anti-missile system of the United States (sometimes called Star Wars). Some models, however, indicate that the introduction of such a defence technique changes the dynamics of the arms race to an extent that chaotic behaviour can occur such that no one can predict how the various parties will react. A similar risk of chaos can arise when we find ourselves in a situation in which many different countries are involved at the same time, each with their own interests and their own grievances, a bit like in the war on terror or the conflicts in the Middle East.

We must be aware that there is insufficient data on the outcome of wars and battles to validate all the models discussed in this chapter. And for good reasons: nobody wants to start a nuclear war to get that data! So, we can ask ourselves the legitimate question of how accurate such models really are. In any case, they can only serve as a guide to get some insight that is not coloured by the emotions of the moment. It would be complete madness to automate such decisions in international politics!

The use of mathematics-based technology in warfare has many other ethical implications. Remote missile control and the use of drones reduce the costs and risks for the party that attacks. The user interface sometimes shows targets as triangles like in a video game, which lowers the threshold to actually shoot. And the general public may get the impression that the weapons are so precise that hardly any innocent civilians are affected. In that situation, it can do no harm to provide sufficient scientific arguments and mathematical models that clearly demonstrate the benefits of cooperation and disarmament. So, there is a great future in research into mathematical models of warfare.

Chapter 20
Pandemics: From Ebola and HIV to Bioterrorism and the Coronavirus

Few events cause panic like the sudden outbreak of a contagious and untreatable disease. In these times of globalization, in which people are becoming increasingly mobile, a local outbreak of a new variant of the coronavirus in Wuhan, China, in 2020 only took a few months to lead to a global pandemic with millions of casualties. In an attempt to contain the damage, governments resorted to severe measures, such as physical distancing, facemasks, limited access to shops, the closing of schools and restaurants. Some countries, such as Spain, Italy, France and Belgium, even imposed lockdowns, in which people were not allowed to leave their homes. But how are such decisions taken? How can policy makers decide what measures will contain a pandemic? And how will they decide when it is safe to relax the constraints?

What's the problem? The coronavirus pandemic that started in 2020 is beyond any doubt one of the most extreme infectious health crises since the development of modern medicine. However, it is certainly not the first time viruses have posed a threat to the world. In 2013, the SARS virus spread from China. In 2014, Ebola broke out in West Africa, and in 2016 the Zika virus was a serious concern emanating from South America. Not only viruses pose problems for global health; more and more bacteria are developing resistance to antibiotics. And also the risk of deliberate contamination by terrorists or hostile states is closely monitored. Since the 1950s, the United States has been developing techniques in its *Centers for Disease Control and Prevention* to assist intelligence services in the face of the threat of a biological attack. An investment that turned out not to be completely paranoid when letters containing the dangerous anthrax bacillus were sent around in the aftermath of the September 11, 2001, attacks. While important, it is very difficult to collect data on the

G. Samaey, J. P. L. Vandewalle, *The Invisible Power of Mathematics*, Copernicus Books, https://doi.org/10.1007/978-1-0716-2776-1_20

likelihood of a large-scale epidemic or its potential impact during normal times, when prevention is the focus. Also during a pandemic, it is not easy to figure out which measures are necessary or sufficient to control disease spreading. In fact: if a pandemic is under control, there is a great tendency to assume that government measures have been too strict, and when the release of measures has dramatic consequences, it is usually too late to turn the situation around. Experiments involving the release of infected individuals into a population are not only impractical, they are also highly unethical. There is only one alternative: we need to look for a mathematical way to model the spread of disease, so that computer simulations can help policy makers take the crucial decisions.

How are diseases transmitted? We can only draw up mathematical models for the transmission of germs when we know exactly how the transmission occurs. For example, SARS is transmitted through the droplets released by coughing or sneezing. This is why infected people wear a face mask: not to protect themselves, but to protect others. The transmission of Ebola requires direct contact with bodily fluids, such as blood, while the Zika virus spreads mainly through infected mosquitoes, rather than from person to person. It is often not immediately clear how exactly a disease is transmitted. For Covid-19, the transmission paths, and their importance, are gradually better understood. This increased understanding systematically leads to an improved accuracy of the mathematical models resulting in better predictions. To unravel these mechanisms, it is often necessary to collect a great deal of data that can reveal links between the infected persons. Moreover, these data must be combined with the correct hypothesis. That this is not obvious is illustrated by an old story: the discovery of the cause of cholera.

The discovery of the cholera bacteria When British physician John Snow investigated the cause of a cholera outbreak in London in 1854, the disease was generally attributed to 'bad air'. Snow did not agree, and suspected that the epidemic was due to the poor quality of London's drinking water. He drew maps on which he indicated water pumps and contaminations, and performed a statistical analysis on these maps. He found that the infections mainly occurred in neighbourhoods where the water pumped up from the Thames had first passed through London. The pumps were located downstream of the city. On this basis, he assumed – without being able to formally demonstrate this – that the drinking water had to be contaminated by human excrement. On his advice, a number of changes were made to the water supply, and the epidemic was curbed. The remarkable feat in this story is that the existence of the cholera bacteria was completely unknown at the time: it took almost 30 years (until 1883) before Robert Koch was able to identify the cholera bacteria. The conclusion that cholera was caused by drinking strongly diluted human excrements – euphemistically referred to in medical jargon as the faeco-oral route – was so distasteful that the scientific world refused to accept it for some time. There have even been cases of people deliberately drinking contaminated water to disprove the theory – an experiment with the odds of survival of Russian roulette.

On to mathematical models As soon as the mechanism of transmission is known, we can start the mathematical modelling work. The easiest way to do this is to consider the probability of infection in case of a risky contact and the number of risky contacts as two separate factors. Let's take the example of HIV, the virus that causes AIDS. To write a mathematical model for the transmission of HIV, we need to know the probability of HIV being transmitted during an unprotected sexual contact. In addition, we need to estimate the number of unprotected sexual contacts. We can then proceed as we did for the prediction of demographic evolution in Chap. 14: we create a large number of *virtual individuals*, which we follow for a certain period of time. For each person, we arrange a number of 'virtual sexual encounters' at a realistic pace, and for each of these encounters there is a chance of transmission if one of the two individuals was infected. This way of modelling is very flexible because we can specify for each individual how often and in what way the sexual encounters take place. For example, we can allow some people to have more sexual contacts with several different people, or indicate that certain forms of sexual contact involve a much greater risk of infection. Of course, this flexibility comes at a certain price: we need a lot of data to make a model of this type accurate enough.

Disease spread through social contacts To determine the frequency with which people have sexual contact, and the number of partners with whom this happens, we rely on surveys in which this question is asked directly. This can lead to inaccurate answers. In such a survey, some will tend to boast and exaggerate their activities, while others will feel a little embarrassed and thus tend to tone down their exploits a little. In any case, there are big differences between individuals. Nevertheless, the reported results show a remarkable mathematical structure, which is best expressed when we present the respondents on a network (such as the social networks from Chap. 2). Each person is then a node, and each sexual relationship a link. In 2001, the Swedish sociologist Fredrik Liljeros showed that the number of sexual partners is statistically distributed according to what is called a 'power law' in the jargon. This probability distribution is shown in Fig. 20.1, and it shows that there is a very large group of people with a limited number of sexual partners, while a very small number of people have a very large number of alternating sexual con-

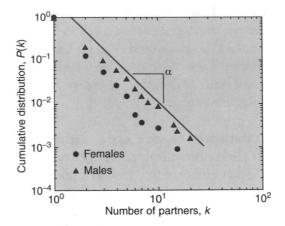

Fig. 20.1 Distribution of the number of sexual partners in the last 12 months over the population as measured in a Swedish survey in 1996. Horizontal axis: number of partners; vertical axis: percentage of the population that reported this number. Note the logarithmic scale on both axes

Fig. 20.2 Simple group model for infectious diseases. People move from the group 'susceptibles' (*S*) to 'infected' (*I*) by contact. They then heal at a fixed rate

tacts. It should come as no surprise that HIV in such a network spreads mainly through those few central hubs with many changing contacts.

The fear of the unknown Measuring the intensity of contact between two people is not always easy. A sexual contact is usually pretty clear, but the frequency of infected handshakes or sneezes is much more difficult to determine. In these circumstances, there is no point in modelling at a very detailed level. To eliminate the unknown factors as much as possible, we usually switch to simpler models with fewer unknown parameters. We then divide the population into groups: the 'susceptibles' (denoted by *S*), the 'infected' (denoted by *I*) and the 'immunes' (denoted by R for 'removed'). Because there is no cure or recovery for HIV, 'immune' means the same as 'deceased' in that case. For most other diseases, 'immune' indicates people who can no longer become ill again. These three groups are shown in Fig. 20.2. We then model the chance at which people pass from one group to another. Infection occurs through contact between an 'infected' and a 'susceptible' person. The probability of such an encounter is proportional to the fraction of susceptible persons S and also to the fraction of infected persons I. Therefore, the rate at which the disease spreads in the population can be written as βSI, with β the probability that an infection takes place during a single encounter. Next, we have the process of recovery, which in normal circumstances only depends on how long a person remains ill. Per unit of time a fraction of γI the sick people recover. The smaller γ, the longer an infected person remains sick and can infect others.

The above schedule is flexible and can be expanded with additional groups. For example, many diseases have an incubation period, in which an infected person is a carrier of the disease, but not yet contagious. This can then give rise to an extra group in the population. Or one can add a group of patients that are hospitalized. These patients then take longer to recover, but also have fewer contacts with susceptible individuals.

Vaccination and group immunity Models like the above one have their limitations, of course. They smooth out all differences between individuals. They therefore do not provide a basis for individuals to assess the consequences of possible risky behaviour. These models no longer take the actual mechanism of disease transmission into account. They can be used for both viruses and bacteria, and models of this kind have even been used to simulate the spread of (extremist) opinions.

It is evident that we can't use such models to precisely predict the future number of infected or hospitalized individuals, but only to get a rough first picture of the risk. This is why such simulation results are always reported with a confidence interval. This confidence interval naturally shows how the uncertainty on predictions grows when looking further in the future. Still, such simple population-based models are very useful, for example, in the development of vaccination schedules. They show, for example, that it is not at all necessary to vaccinate the entire population in order to prevent epidemics. It is sufficient to make the group of vaccinated people so large that on average each infected person can infect less than one other. When that is the case, the vaccinated part of the population automatically protects the others, an effect we call group or herd immunity. The critical fraction of vaccinated people depends on the disease, and more specifically on the balance between the duration of the infection and the ease with which it is passed on. For example, for flu, it is sufficient to vaccinate about 40% of the population, while for measles, a vaccination coverage of 95% is required. That is why the latter is compulsory.

Critical vaccination coverage and polio eradication

To avoid an epidemic, it is important that an infection does not spread too quickly. As shown in the group model in Fig. 20.2, the number of new infections per time unit is equal to βSI, while the number of cures per time unit is given by γI. The total number of sick people increases when there are more new infections per unit of time than people that recover. That is the case when, $\beta SI > \gamma I$, or in other words, when

$$(\beta S - \gamma)I > 0.$$

There is some logic to this comparison: the more contagious the disease (higher β), the quicker an epidemic develops. The faster an infected person recovers (higher γ), the slower. Unfortunately, these two factors are out of our control because they depend on the disease itself. For example, measles are more contagious than the flu, and people also recover less quickly from measles than from the flu. There is actually not much that can be done about those two factors. The only element in the equation that we have under control is the fraction of 'susceptibles'S. That is a number that we can get down by vaccinating. If we assume that there are currently no infected people, the fraction of 'susceptible people' equals 100%, i.e. 1. If we now vaccinate a fraction V of the population, the fraction of 'susceptibles' will become $S = 1 - V$. It is now easy to calculate what fraction of vaccinated people is needed to prevent the epidemic: we need $\beta(1 - V) - \gamma < 0$ or $V > 1 - \gamma/\beta$. For measles we know that $\beta = 24$ and $\gamma = 1/7$ (because recovery takes on average 7 days). This means that $V > 0.94$ and a vaccination coverage of more than 94% is needed to prevent an epidemic of measles.

To achieve herd immunity against polio, a vaccination coverage of 85% is required. One can only fully assess how spectacular the global eradication of polio really is when realizing this. Herd immunity is fragile. For instance, in the Netherlands, certain orthodox-protestant groups do not allow themselves to be vaccinated for religious reasons: they believe that they enjoy divine protection. However, from the above discussion, we can only conclude that this divine protection only comes to them through the vast majority of vaccinated 'non-believers' among whom they live.

Biological warfare and bioterrorism Biological warfare is probably as old as war itself. The oldest known writings on the subject date from 2000 to 1500 B.C., and speak of hare plague victims driven to hostile territory by the Hittites (a population group in Anatolia, present-day Turkey) to cause an epidemic there. Also the original population of America was massively exterminated by the European settlers spreading the smallpox virus to which they themselves were immune. Usually this happened by accident, but there are documented cases of deliberate spread. A systematic development of biological weapons was to wait until the second half of the nineteenth century, when

Koch was able to isolate bacteria as disease carriers. Shortly after that discovery, great efforts were made to transform bacteria into effective biological weapons. For example, during the First World War, Germany experimented with anthrax to infect cattle sold by neutral countries to the Allies. During the Cold War, all military superpowers developed extensive biological warfare programs.

The role of the friendship paradox In biological warfare, it is not only important to develop new biological attack weapons. Ways to 'deliver' them efficiently are also crucial. And of course, there is also the defensive component: treatments and vaccination strategies. For example, in the case of an attack with an unprecedented biological agent, it is often impossible to have sufficient vaccines in stock to vaccinate everyone quickly. Often a vaccine still needs to be developed. And once it has been developed, it usually cannot be produced immediately in large quantities. An important question then is: when we only have a limited amount of vaccines available, who do we preferably vaccinate? A question that becomes particularly relevant at a time when no one dares to rule out the possibility that biological weapons will fall into the hands of a group of terrorists with an apocalyptic world view.

This is where our analysis of social networks from Chap. 2 can come in handy, especially the friendship paradox, which concluded that on average the friends of a random person have more friends than that random person himself. That fact can be used to vaccinate much more efficiently than the entire population. The theory above shows that an aggressive disease can only be stopped effectively by vaccinating random people when we vaccinate 90–95% of the population. By using small world networks, we can show that a vaccination of 20–40% of the population would suffice when we vaccinate *random friends of random people*. We should put a caveat here. During the Covid-19 crisis that started in 2020, such a strategy was not used, despite the relatively slow production of vaccines. This is due to a very important aspect that we have not treated in this chapter: Covid-19 does not have the same health consequences for every susceptible individual: old people are much more likely to die than young people. In that case, it is evidently more important to protect the most vulnerable than to gamble on a slightly earlier herd immunity.

When it comes to the spread of diseases, mathematics plays a crucial role. In addition to the emergence of epidemics and the study of vaccination schedules, there are many questions that mathematics can help with. Detecting that an epidemic is going on, for example. At what point is an increasing number of infections alarming? Until when is it an accidental fluctuation? Drastic measures are often needed to completely eradicate an epidemic, as the Ebola crisis of 2014 shows. These measures can cause panic and major economic damage. So it is best not to use these unnecessarily. Reacting too late is, of course, not an option either. To tackle this problem, statistical methods are being developed that detect when an increase in the number of infections can no longer be explained by chance. Also, during the Covid-19 pandemic, mathematical models have been used to calculate predictions for various scenarios that advise governments on alternative paths to gradually re-open society. It is then up to policy makers to weigh these predictions along with other advice. We cannot add much more evidence that mathematics can be of vital importance.

Bibliographic References

Charles A. Dana Center at The University of Texas at Austin, *"Inside mathematics"*, http://www.insidemathematics.org

ERCIM Newsletter, *"Introduction to the special theme: Maths for Everyday Life"* http://ercim-news.ercim.eu/en73/special/introduction-to-the-special-theme-maths-for-everyday-life

National Academy of Sciences, *"Fueling Innovation and Discovery: The Mathematical Sciences in the 21st Century"*, http://www.nap.edu/catalog/13373/fueling-innovation-and-discovery-the-mathematical-sciences-in-the-21st

Mols, B., Smeets, I., *"Succesformules, Toepassingen van wiskunde"*, Platform Wiskunde Nederland PWN. http://www.platformwiskunde.nl/files/documenten/PWNsuccesformules_LowRes.pdf

Stichting Math4All, http://www.math4all.nl

"Mathematical sciences and their value for the Dutch economy", Platform Wiskunde Nederland, Deloitte, 2014, http://www.euro-math-soc.eu/system/files/uploads/DeloitteNL.pdf

Lery, T., Primicerio, M. e.a. (eds), *"European success stories in industrial mathematics"*, Springer, 2011.

Deuflhard, P., Grötschel, M. e.a. (eds), *"Matheon – mathematics for key technologies"* European Mathematical Society, Publishing House, Zurich, EMS Series Industrial and Applied Mathematics, Vol. 1.

Mahajan, S., *"Streetfighting mathematics, The Art of Educated Guessing and Opportunistic Problem Solving"*, MIT Press, 2010.

Wigner, E., *"The unreasonable effectiveness of mathematics in the natural sciences"*, Communications in Pure and Applied Mathematics, John Wiley & Sons, Vol.13, No 1, 1960.

Hamming, R. " *The unreasonable effectiveness of mathematics*", American Mathematics Monthly, Wiley, 87, 81–90 1980.

Schools plus competition, "*Where would the world be without mathematics ?*", http://www.bshm.ac.uk/plus

Rousseau, C., Saint-Aubin, Y, "*Mathematics and Technology*", Springer 2008.

"*Mathematical Moments*", American Mathematical Society, AMS, http://www.ams.org/samplings/mathmoments/mathmoments

"*Why do math*", Society for Industrial and Applied Mathematics, SIAM 2011, http://www.whydomath.org

"SIAM Nuggets", Society for Industrial and Applied Mathematics, SIAM http://connect.siam.org/category/siam-nuggets/

"*Grand Challenges for Engineering,*" National Academy of Engineering of the National Academies, Washington, USA. http://www.engineeringchallenges.org

Junior College KU Leuven "*Van priemgetal tot digitale handtekening*" https://www.kuleuven.be/onderwijs/juniorcollege/STEM/themas#4

Watts, D.J., Strogatz, S.H.,"Collective dynamics of 'small-world' networks.", Nature, Macmillan Publishers, 1998 June 4;393(6684):440–2.

"Six degrees of separation" toneelstuk en YouTube, https://en.wikipedia.org/wiki/Six_Degrees_of_Separation_(play),https://www.youtube.com/watch?v=TcxZSmzPw8k

Junior College KU Leuven, "*Cochleaire implantaten: wat is geluid, de werking van het oor, signaal- en Fourieranalyse, de werking van een implantaat*" http://set.kuleuven.be/English/news/2014/junior-college-stem

M.D. Charles Limb TED-talk: "*Building the musical muscle*", TED Talks, 2011, http://www.ted.com/talks/charles_limb_building_the_musical_muscle.html

Nguyen, T., Zupancic, S., Lee, D.Y.C. "*Engineering challenges in cochlear implants design and practice*", IEEE Circuits and Systems Magazine, Fourth Quarter, 2012 pp. 47–55.

Varshney, L.R., Sun, J.Z., "*Why do we perceive logarithmically?*" Significance, The Royal Statistical Society, Wiley, February 2013 pp.28–31

Junior College, KU Leuven: "*PageRank, de kracht van Google: een toevallige surfer op het web, eigenvectoren en eigenwaarden, benaderen van eigenvectoren, enkele statistische aspecten*" https://www.kuleuven.be/onderwijs/juniorcollege/STEM/themas#3

Groote, J. F., Zantema H., "*A probabilistic analysis of the Game of the Goose*", Computer Science Reports 14-04, Eindhoven, 2014 http://purl.tue.nl/688639272109161

Demeulenaere, B., Coemelck, D., Hemelsoen, J., Aertbeliën, E., Verschuure, M., Roelstraete, K., Swevers, J., De Schutter, J. "*Optimaal balanceren voor weven aan onvergelijkbare snelheden*", Het Ingenieursblad, Antwerpen, 77 (2), 23–27, 2008.

Peeters, B., Climent, H., e.a. "*Modern solutions for Ground Vibration Testing of large aircraft*" Proceedings of IMAC 26, International Modal Analysis Conference, Orlando (FL), USA, 4–7 February 2008.

Van Der Auweraer, H., Anthonis, J., De Bruyne, S., Leuridan, J. (2012). "*Virtual engineering at work: The challenges for designing mechatronic products*", Engineering With Computers, 29 (3), 389–408.

Wijnveld, F. *"Safety in football stadiums"*, Crowd Management, http://crowdprofessionals.nl/crowd-management-2/safety-football-stadiums-2.html

Sipics, M., *"The Van Gogh Project: Art Meets Mathematics in Ongoing International Study"*, SIAM News, May 18, 2009.

Wald, C. *"Wheels when you need them"*, Science, AAAS, 22 august, 2014, Vol345, issue 6199.

Glover, F., Sörensen K., *"Metaheuristics"*, Scholarpedia, 10(4):6532, 2015.

Rainer-Harbach, M., Papazek, P., e.a., *"PILOT, GRASP, and VNS approaches for the static balancing of bicycle sharing systems, "* Journal of Global Optimization, Springer, November 2015, Volume 63, Issue 3, pp 597–629.

Robert Bridson, *Mathematician and Oscar winner.* http://www.cs.ubc.ca/~rbridson/

Steiner, C. *"Automate this"*, Penguin, UK, 2012.

Bar-Natan, D., *"Dessert Hilbert's 13th problem, in full color"*, Nov. 2009. https://www.math.toronto.edu/drorbn/Talks/Fields-0911/Hilbert13th.pdf

Jain, A., Mao, J., Mohiuddin, *"Artificial Neural Networks: A Tutorial "*, IEEE Computer, Issue No.3, March, 1996 vol.29, pp. 31–44.

Van Gestel T., Suykens J.A.K., e.a. ``*Financial Time Series Prediction using Least Squares Support Vector Machines within the Evidence Framework*'', IEEE Transactions on Neural Networks, vol. 12, no. 4, Jul. 2001, pp. 809–821.

Espinoza M., Suykens J.A.K., Belmans R., De Moor B., ``*Electric Load Forecasting - Using kernel based modeling for nonlinear system identification*'', IEEE Control Systems Magazine, vol. 27, no. 5, Oct. 2007, pp. 43–57.

Weigend A.E., Gerschenfeld N. A., (eds), *"Time series prediction; Forecasting the future and understanding the past"*, Addison Wesley, Santa Fe Institute, 1994.

McNames J., Suykens J.A.K., Vandewalle J., ``*Winning Entry of the K.U.Leuven Time-Series Prediction Competition*'', International Journal of Bifurcation and Chaos, vol. 9, no. 8, Aug. 1999, pp. 1485–1500.

Grow, A., Van Bavel, A., *"Assortative Mating and the Reversal of Gender Inequality in Education in Europe: An Agent-Based Model "*, PLOS one, Open Access, June 3, 2015.

Stix, G., *"Profile: David A. Huffman, Encoding the "Neatness" of Ones and Zeroes"*, Scientific American, pp. 54–58, September 1991.

CITYNETMOBIL Result In Brief, *"Driverless cars take to the road"*, http://cordis.europa.eu/result/rcn/90263_en.html

Case, J., *"Understanding systemic risk in financial markets "*, SIAM News, Philadelphia, Vol.45, No 10, Dec.2012.

May, R., Levin, S., Sugihara, S., *"Complex systems: Ecology for bankers"*, Nature, Macmillan Publishers, 451, 893–895 20 February 2008.

Junior College KU Leuven, *"Simulatie van de klimaatverandering. Differentiaalvergelijkingen, modellering, stabiliteit en numerieke methodes."* https://www.kuleuven.be/onderwijs/juniorcollege/STEM/themas#4

Kaper, H., Engler, H., *"Mathematics and Climate"*, SIAM books, Philadelphia, Oct. 2013. ISBN: 9781611972603

Mackenzie, D., *"Mathematics of Climate Change, A new discipline for an uncertain century"*, Mathematical Sciences Research Institute, MRSI, Berkeley, CA, 2007.

Randall, D., Wood, R., e.a. "Climate models and their evaluation", IPCC, Chapter 8, in Climate Change 2007, Cambridge University Press, Cambridge UK.

"Wat zijn de oorzaken en de gevolgen van klimaatverandering ?" Koninklijke Nederlandse Academie van Wetenschappen, https://www.knaw.nl/nl/thematisch/de-nederlandse-wetenschapsagenda/aarde-klimaat-energie-en-bio-omgeving/wat-zijn-oorzaken-en-gevolgen-klimaatverandering

Booß-Bavnbek, B., Høyrup, J. (eds), "Mathematics and War", Springer, 2003, ISBN: 978-3-7643-1634-1

Index

A

Acoustic radiation, 72
Aerodynamics, 70
Aeroelastic flutter, 71
Aging, 79
Airplane wings, 72, 73
Algorithms, 4–6, 18, 19, 35, 39, 40, 45,
        64, 84, 85, 87, 98, 99, 105, 107,
        122, 125, 126, 131, 148, 149
Alliances, 17, 18
Arithmetic mean, 115
Arms races, 153, 155–157
Art historians, ix, 81
Artificial brains, 97
Artificial neural networks, 95–99, 105,
        107, 131
ASCII code, 120–122
Attractions, 70, 83, 113–115
Auditory nerves, 31–33, 35
Authenticity, ix, 7, 9, 79
Avatar, 89

B

Balancing car wheels, 61
Banking systems, 137–144

Bias, 26, 27, 29
Bike sharing, 85
Bike stations, 84, 85
Biometrics, 119
Bioterrorism, 160–166
Bitcoin, 9
Black box models, 104, 105
Block diagram, 132, 133
Board games, ix, 47
Bridges, 69–75, 148
Butterfly effect, 102, 103

C

Calculus, 72
Cameras, 38, 81, 99, 126, 130
Car sharing, 134
Chaotic behaviour, 105, 157
Checkers, 49, 50
Checksum, 123
Chess, 49
Cholera bacteria, 160, 161
Cilia, 31, 32
Cinema, v, 78
Citadels, 50
Clay prize, x

Climate, 55, 60, 142, 144–151
Climate panel IPCC, 149
Clouds, 3, 89, 148
CO_2 amount, 147
Cochlea, 31–33
Cochlear implants, 31–38, 124
Codes, 4, 6, 8, 122–124
Complexity theory, 84
Compression, v, 122–126
Computational mathematics, 149
Computer animation, 89
Computer controls, 9
Computer fraud, 91–99
Computer programs, 6, 22, 23, 39,
    57, 96, 135
Computer simulations, 58, 59, 89,
    113, 150, 160
Computer virus, 126
Computing time, 35
Confidence interval, 24, 163
Controlled disarmament treaties, 155
Controller, 133
Convex, 64, 65
Cooley and Tukey's algorithm, 35
Cosines, 33–35
Crash tests, 74
Credit card companies, 94
Cruise control, 132

D

Data, 4, 7, 8, 14, 45, 91, 92, 97,
    103–107, 111, 114, 116,
    119–127, 144, 149, 151, 155,
    157, 159–161
Decibels, 37, 38
Demographic processes, 111, 115
Derivatives, 64, 137, 140, 147
Design criteria, 57
Diagnoses, 91–93, 98, 125
Differential equations, 147, 156
Digipasses, 7, 8
Digital restoration, 80, 81
Digital watermarks, 78

Disaster scenarios, 69
Dispersal of marine debris, 151
Dividing line, 6, 29
Driving comfort, 62, 66
Dynamics, 66, 143, 146, 154, 157

E

Eardrum, 31
Ebola virus, 159–166
Ecosystems, 143, 144
Eigenvalues, 43
Electricity usage, 56, 59, 104
Electrodes, 33, 35, 36
Electronic payments, 9
Encrypts, 4
Engine noise, 66
Engineers, ix, xi, 36, 70, 73, 103
Error correction and detection, 123
Experimental psychology, 37
Exponential complexity, 84, 85

F

Facebook, 11, 12, 19, 99, 119
Factory noise, 61–67
Fast Fourier transform, 35
Fermat's theorem, 6
Fields Medal, x, 103
Files, 123, 124
Finite elements, 72, 73
Flash crashes, 139, 140
Flemish industry, 61
Football spectators, 74
Football stadiums, 69–75
Forces, 57, 65, 71–74, 81,
    114, 135
Forgery of painting, 77
Forward contracts, 137
Four in a row, 51
Fourier transform, 33–35
Fractions, x, 14–16, 24, 42–44, 146,
    147, 163, 164
Fraudulent transactions, 97, 99

Frequency and amplitude, 33
Fresh crawl, 44
Friendship connections, 16
Friendship paradox, 12, 165
Fuel consumption, 62, 129, 132, 134

G

Game theory, 47–52
Genetic algorithms, 86, 87
Geometric mean, 115, 116
Global minimum, 64
Global warming, ix
Google dance, 44
Google search, 44
Googol, 120
Goose board, 47
Gravity movie, 89

H

Hearing aids, ix, 31, 33, 35
Hearing damage, 61
Heat balance of the earth, 146
Heuristic methods, 86
High-dimensional space, 95
Hilbert 13 problem, 96

I

Identity, 4, 6–9, 127
Ig-Nobel Prize, 20
Image processing, 79
Imbalances, 27, 62, 83
Impossibility theorem of Balinski and
     Young, 28
Industrial revolution, 55
Infections, 160, 161, 163–166
Information theory, 154
Insurances, 135, 137
International security, 156
Internet, 3–5, 9, 40–42, 44, 77, 78,
     119, 122, 127
Inverse problems, 151

K

Keller, Joe, 20

L

Like, 12
LinkedIn, 11
Loudness, 31

M

Machine learning, 105, 106
Magnetic radiation from mobile
     phone, 75
Mass production, 75
Mathematical engineering
     techniques, xi
Mathematical models, 16, 58, 59, 73,
     74, 89, 101–105, 111,
     115–117, 130, 137–139, 144,
     145, 149, 151, 155, 157, 160,
     161, 166
Messages, 4–7, 12, 13, 29, 102, 122,
     123, 153
Milgram experiment, 14
Military technology, 154
Mobility, ix, 129, 135
Modelling technique, 105, 111,
     143, 145
Molecules, 112, 113
Morse code, 123
Mortgage loan, 140
Movie scene, 125
MP3 compression, 124
Mutation, 86
Mystic Lamb of God, 77

N

Nash equilibrium, 51, 52
Natural frequencies, 32, 71, 72
Networks, 7, 9, 11–19, 41, 42, 47, 59,
     60, 95–97, 104, 114, 143,
     161, 162, 165

Neurons, 19, 95–97, 107
Nonlinear functions, 96, 107
Nuclear chain reactions, 153

O

Ocean currents, 149, 151
Online dating, 115, 116
Online shopping, ix, 3, 9
Operational research, 154
Opinion polls, ix, 21–27, 29
Optimization, 58, 61, 62, 64–66, 88
Options, 27, 74, 107, 135, 138, 153,
        154, 156, 166
Ossicles, 31

P

Pagerank, 39–45
Paradoxes, 27, 28
Parameters, 15, 63, 92, 95, 114–116,
        139, 147, 148, 154, 155, 162
Parliamentary elections, 22, 28
Pascal, 37, 48, 49
Passwords, 4, 7, 8
Pattern recognition, 91
Perception, 38
Perelman, Grigori, x
Personality types, 98
Phishing, 4
Physical principles models, 102, 105
Picture Archiving and Communication
        System (PACS), 126
Pixels, 77–81, 125
P *vs*. NP, 85
Policy makers, 159, 160, 166
Polls, 21–29
Ponytails, 20
Power grid, 55, 59, 60, 101
Prime factors, 6
Privacy, 6, 99
Problem size, 84, 85

Professional career, 117
Professional life, ix
Prosthesis, 127
Protocol, 4, 7
Public debate, 9
Public key, v, 4–7
Public opinion, x, 28
Public transport, 83, 88

Q

Quotient series, 28

R

Reach in the world, 11–20
Reactor vessel, 112
Relatively, 17, 65, 84, 85, 112, 131,
        139, 143, 165
Renewable energy, ix, 55, 59, 60, 106
Reproduction, 36, 86
Resonance, 71
Reversal of large conveyor belt, 149
Risks, vii, 9, 50, 59, 89, 99, 101,
        137–141, 143, 154, 155, 157,
        159, 161, 163
Robustness, 108, 143
Rounding mechanisms, 21
Routing, 84–87
RSA algorithm, 5
Rubber ducks, 151

S

Safety, 57, 58, 62, 66, 69, 74, 101,
        129, 131, 133
Samples, 21, 23–27
Search engines, 39, 40, 44
Self-driving car, 129, 131–134
Sensors, 38, 59, 66, 74, 119, 126, 130,
        134, 135
Simulate, 104, 163

Sines, 33–35
Single, 24, 33, 40, 45, 57, 63, 64, 74,
    80, 111–117, 120, 121, 163
Small world networking, 15,
    16, 19, 165
Social media, 11, 13
Sound masking, 125
Spy program PRISM, 7
Stability of drones, 75
Star Wars, 157
Statistical errors, 23, 26, 29
Statistics and probability, 23
Stelarc, 127
Stochastic processes, 153
Strategies, 47, 50–52, 154, 155, 165
Strategy, 138, 141
Supercomputer, 73, 148, 149
Surfers, 40–44
Synapses, 95–97
Synthetic sound, 66
Systemic risks, 140, 141, 143, 144
Systems, 3, 8, 9, 21, 22, 28, 33, 39, 66,
    83, 87, 102, 107, 113, 117,
    126, 127, 132–134, 137,
    139–144, 149, 154, 156, 157

T

Tacoma Narrows Bridge, 70, 71
Teamwork, 103, 104
Teapots, 20
Time series, 104, 106, 107
Timetables, 88
To drive, 130–135
To generate electricity, 57

To predict (Final stage), 101
Turbulence, 58, 59
Twitter, 11, 19

U

Underlying laws, 105

V

Van Goghs, 79
Vibrations, 31, 33, 34, 61–65,
    71–73, 75
Virtual, 9, 22–25, 58, 69–75, 114, 161
Virtual prototypes, 73–75
Volatility, 139
Vulnerable, 137–144, 165

W

War and peace, 153–157
Watts-Strogatz network, 16
Wavelets, 79–81
Weather forecasts, 106, 155
Weaving machines, 61, 62, 64, 65
Weber-Fechner's law, 37
Webpages, 44
Wind tunnels, 58, 70
Wind turbines, 55–59
Windmills, 56, 57
World peace, ix

Z

Zip, 124

Printed in the United States
by Baker & Taylor Publisher Services